T0313712

An Introduction to Self-Adaptive Systems

A Contemporary Software Engineering Perspective

Danny Weyns
Katholieke Universiteit Leuven, Belgium

Registered Offices
John Wiley & Sons, Inc., 111 River Street, Hoboken, NJ 07030, USA
John Wiley & Sons Ltd, The Atrium, Southern Gate, Chichester, West Sussex, PO19 8SQ, UK

Editorial Office
The Atrium, Southern Gate, Chichester, West Sussex, PO19 8SQ, UK

For details of our global editorial offices, customer services, and more information about Wiley products visit us at www.wiley.com.

Wiley also publishes its books in a variety of electronic formats and by print-on-demand. Some content that appears in standard print versions of this book may not be available in other formats.

Library of Congress Cataloging-in-Publication Data applied for,

Hardback ISBN: 9781119574941

Cover Design: Wiley
Cover Image: © Takeshi.K/Getty Images

Set in 9.5/12.5pt STIXTwoText by SPi Global, Chennai, India

Printed and bound by CPI Group (UK) Ltd, Croydon, CR0 4YY

10 9 8 7 6 5 4 3 2 1

To Frankie

v

Contents

Foreword *xi*
Acknowledgments *xv*
Acronyms *xvii*
Introduction *xix*

1   **Basic Principles of Self-Adaptation and Conceptual Model** *1*
1.1   Principles of Self-Adaptation *2*
1.2   Other Adaptation Approaches *4*
1.3   Scope of Self-Adaptation *5*
1.4   Conceptual Model of a Self-Adaptive System *5*
1.4.1   Environment *5*
1.4.2   Managed System *7*
1.4.3   Adaptation Goals *8*
1.4.4   Feedback Loop *8*
1.4.5   Conceptual Model Applied *10*
1.5   A Note on Model Abstractions *11*
1.6   Summary *11*
1.7   Exercises *12*
1.8   Bibliographic Notes *14*

2   **Engineering Self-Adaptive Systems: A Short Tour in Seven Waves** *17*
2.1   Overview of the Waves *18*
2.2   Contributions Enabled by the Waves *20*
2.3   Waves Over Time with Selected Work *20*
2.4   Summary *22*
2.5   Bibliographic Notes *23*

3   **Internet-of-Things Application** *25*
3.1   Technical Description *25*
3.2   Uncertainties *28*
3.3   Quality Requirements and Adaptation Problem *29*
3.4   Summary *29*

3.5   Exercises   *30*
3.6   Bibliographic Notes   *31*

4   **Wave I: Automating Tasks**   *33*
4.1   Autonomic Computing   *34*
4.2   Utility Functions   *35*
4.3   Essential Maintenance Tasks for Automation   *37*
4.3.1   Self-Optimization   *37*
4.3.2   Self-Healing   *38*
4.3.3   Self-Protection   *40*
4.3.4   Self-Configuration   *42*
4.4   Primary Functions of Self-Adaptation   *43*
4.4.1   Knowledge   *44*
4.4.2   Monitor   *46*
4.4.3   Analyzer   *47*
4.4.4   Planner   *49*
4.4.5   Executor   *51*
4.5   Software Evolution and Self-Adaptation   *52*
4.5.1   Software Evolution Management   *53*
4.5.2   Self-Adaptation Management   *54*
4.5.3   Integrating Software Evolution and Self-Adaptation   *55*
4.6   Summary   *56*
4.7   Exercises   *59*
4.8   Bibliographic Notes   *60*

5   **Wave II: Architecture-based Adaptation**   *63*
5.1   Rationale for an Architectural Perspective   *64*
5.2   Three-Layer Model for Self-Adaptive Systems   *66*
5.2.1   Component Control   *67*
5.2.2   Change Management   *67*
5.2.3   Goal Management   *68*
5.2.4   Three-Layer Model Applied to DeltaIoT   *68*
5.2.5   Mapping Between the Three-Layer Model and the Conceptual Model for Self-Adaptation   *70*
5.3   Reasoning about Adaptation using an Architectural Model   *70*
5.3.1   Runtime Architecture of Architecture-based Adaptation   *71*
5.3.2   Architecture-based Adaptation of the Web-based Client-Server System   *73*
5.4   Comprehensive Reference Model for Self-Adaptation   *75*
5.4.1   Reflection Perspective on Self-Adaptation   *76*
5.4.2   MAPE-K Perspective on Self-Adaptation   *78*
5.4.3   Distribution Perspective on Self-Adaptation   *79*
5.5   Summary   *83*
5.6   Exercises   *84*
5.7   Bibliographic Notes   *87*

6   **Wave III: Runtime Models**   *89*
6.1   What is a Runtime Model?   *90*
6.2   Causality and Weak Causality   *90*
6.3   Motivations for Runtime Models   *91*
6.4   Dimensions of Runtime Models   *92*
6.4.1   Structural versus Behavioral   *93*
6.4.2   Declarative versus Procedural   *94*
6.4.3   Functional versus Qualitative   *95*
6.4.3.1   Functional Models   *95*
6.4.3.2   Quality Models   *95*
6.4.4   Formal versus Informal   *98*
6.5   Principal Strategies for Using Runtime Models   *101*
6.5.1   MAPE Components Share K Models   *101*
6.5.2   MAPE Components Exchange K Models   *103*
6.5.2.1   Runtime Models   *103*
6.5.2.2   Components of the Managing System   *104*
6.5.3   MAPE Models Share K Models   *105*
6.6   Summary   *108*
6.7   Exercises   *109*
6.8   Bibliographic Notes   *114*

7   **Wave IV: Requirements-driven Adaptation**   *115*
7.1   Relaxing Requirements for Self-Adaptation   *116*
7.1.1   Specification Language to Relax Requirements   *116*
7.1.1.1   Language Operators for Handling Uncertainty   *116*
7.1.1.2   Semantics of Language Primitives   *118*
7.1.2   Operationalization of Relaxed Requirements   *118*
7.1.2.1   Handing Uncertainty   *118*
7.1.2.2   Requirements Reflection and Mitigation Mechanisms   *119*
7.1.2.3   A Note on the Realization of Requirements Reflection   *121*
7.2   Meta-Requirements for Self-Adaptation   *122*
7.2.1   Awareness Requirements   *123*
7.2.2   Evolution Requirements   *124*
7.2.3   Operationalization of Meta-requirements   *126*
7.3   Functional Requirements of Feedback Loops   *127*
7.3.1   Design and Verify Feedback Loop Model   *128*
7.3.2   Deploy and Execute Verified Feedback Loop Model   *130*
7.4   Summary   *131*
7.5   Exercises   *132*
7.6   Bibliographic Notes   *134*

8   **Wave V: Guarantees Under Uncertainties**   *137*
8.1   Uncertainties in Self-Adaptive Systems   *139*
8.2   Taming Uncertainty with Formal Techniques   *141*
8.2.1   Analysis of Adaptation Options   *141*

8.2.2   Selection of Best Adaptation Option   *143*
8.3   Exhaustive Verification to Provide Guarantees for Adaptation Goals   *144*
8.4   Statistical Verification to Provide Guarantees for Adaptation Goals   *149*
8.5   Proactive Decision-Making using Probabilistic Model Checking   *154*
8.6   A Note on Verification and Validation   *160*
8.7   Integrated Process to Tame Uncertainty   *160*
8.7.1   Stage I: Implement and Verify the Managing System   *161*
8.7.2   Stage II: Deploy the Managing System   *162*
8.7.3   Stage III: Verify Adaptation Options, Decide, and Adapt   *163*
8.7.4   Stage IV: Evolve Adaptation Goals and Managing System   *163*
8.8   Summary   *164*
8.9   Exercises   *165*
8.10   Bibliographic Notes   *168*

9   Wave VI: Control-based Software Adaptation   *171*
9.1   A Brief Introduction to Control Theory   *173*
9.1.1   Controller Design   *174*
9.1.2   Control Properties   *175*
9.1.3   SISO and MIMO Control Systems   *176*
9.1.4   Adaptive Control   *177*
9.2   Automatic Construction of SISO Controllers   *177*
9.2.1   Phases of Controller Construction and Operation   *178*
9.2.2   Model Updates   *179*
9.2.3   Formal Guarantees   *181*
9.2.4   Example: Geo-Localization Service   *183*
9.3   Automatic Construction of MIMO Controllers   *184*
9.3.1   Phases of Controller Construction and Operation   *184*
9.3.2   Formal Guarantees   *186*
9.3.3   Example: Unmanned Underwater Vehicle   *186*
9.4   Model Predictive Control   *189*
9.4.1   Controller Construction and Operation   *189*
9.4.2   Formal Assessment   *191*
9.4.3   Example: Video Compression   *192*
9.5   A Note on Control Guarantees   *194*
9.6   Summary   *194*
9.7   Exercises   *196*
9.8   Bibliographic Notes   *199*

10   Wave VII: Learning from Experience   *201*
10.1   Keeping Runtime Models Up-to-Date Using Learning   *203*
10.1.1   Runtime Quality Model   *204*
10.1.2   Overview of Bayesian Approach   *205*
10.2   Reducing Large Adaptation Spaces Using Learning   *208*
10.2.1   Illustration of the Problem   *208*
10.2.2   Overview of the Learning Approach   *210*

10.3   Learning and Improving Scaling Rules of a Cloud Infrastructure   *213*
10.3.1   Overview of the Fuzzy Learning Approach   *214*
10.3.1.1   Fuzzy Logic Controller   *214*
10.3.1.2   Fuzzy Q-learning   *217*
10.3.1.3   Experiments   *221*
10.4   Summary   *223*
10.5   Exercises   *225*
10.6   Bibliographic Notes   *226*

11   Maturity of the Field and Open Challenges   *227*
11.1   Analysis of the Maturity of the Field   *227*
11.1.1   Basic Research   *227*
11.1.2   Concept Formulation   *228*
11.1.3   Development and Extension   *229*
11.1.4   Internal Enhancement and Exploration   *229*
11.1.5   External Enhancement and Exploration   *230*
11.1.6   Popularization   *230*
11.1.7   Conclusion   *231*
11.2   Open Challenges   *231*
11.2.1   Challenges Within the Current Waves   *231*
11.2.1.1   Evidence for the Value of Self-Adaptation   *231*
11.2.1.2   Decentralized Settings   *232*
11.2.1.3   Domain-Specific Modeling Languages   *232*
11.2.1.4   Changing Goals at Runtime   *233*
11.2.1.5   Complex Types of Uncertainties   *233*
11.2.1.6   Control Properties versus Quality Properties   *234*
11.2.1.7   Search-based Techniques   *234*
11.2.2   Challenges Beyond the Current Waves   *235*
11.2.2.1   Exploiting Artificial Intelligence   *235*
11.2.2.2   Dealing with Unanticipated Change   *236*
11.2.2.3   Trust and Humans in the Loop   *236*
11.2.2.4   Ethics for Self-Adaptive Systems   *237*
11.3   Epilogue   *239*

Bibliography   *241*
Index   *263*

Foreword

From the earliest days of computing, theorists recognized that one of the most striking aspects of computation is its potential ability to change itself: rather than presenting users with a fixed set of computations determined at deployment, the system could at runtime modify both what it computes and how it computes it. However, while "self-modification" was perhaps interesting from a theoretical point of view, few programming systems and engineering methods embraced this capability – the advantages of doing so were not obvious given the additional complexity of reasoning about system behavior and the potential for inadvertently making a really big mess of things.

Over the past decade, however, self-adaptive systems have emerged as a fundamental element of modern software systems. Virtually all enterprise systems have built-in adaptive mechanisms to handle faults, resource management, and attacks. Increasingly systems are taking over tasks otherwise performed by humans in transportation (automated driving), medicine (assisted diagnosis), environmental control (smart buildings), and many others.

In the broad field of software engineering these changes have been mirrored in a number of seismic shifts. The first has been a shift in focus from development time to runtime. For most of its history, software engineering primarily focused on getting things "right" before a system was deployed. When problems were encountered, systems were taken off-line and fixed before redeployment. This made sense because few systems required non-stop availability and hence they could be taken down for "scheduled maintenance." But today almost all public facing systems must be continuously available, requiring that systems be modifiable (either automatically or by developers) while they continue to operate.

A second shift has been the increasing level of uncertainty that accompanies modern systems. In the past, software was typically developed for a known environment using components that were largely under the control of the developers. Today systems work in much more uncertain contexts: loads can change dramatically; resources (such as network bandwidth) can vary substantially for mobile computing; faults can arise from interaction with other systems and resources outside the control of the developer; and attacks can emerge in unexpected ways.

A third shift has been an interest in automation to reduce the cost of operations. In the 1980s it was recognized that while the cost of acquiring or developing increasingly complex computing systems was steadily declining, the overall cost of ownership was rising. The reason for this was the need for system administration, which was taking up a larger and larger fraction of the IT operational budget. By automating many of the routine functions performed by administrators, systems would become more affordable. Moreover, arguably, for today's complex software systems, complete human oversight and control is simply not possible.

A fourth shift has been the commoditization of AI. Whereas for much of its existence AI had largely been relegated to special niches (e.g., robotics), the increasing availability of planners, machine learning, genetic algorithms, and game-theoretic decision systems has made it possible to harness sophisticated reasoning and learning mechanisms in support of automation.

All of these shifts have led to a set of critical challenges for software engineers. What are the fundamental unifying principles that underlie self-adaptive systems? How should one go about engineering them in a way that allows us to assure the system matches its requirements even as those requirements change after deployment? How can we provide safeguards against adaptation-gone-awry? How do we engender trust in systems where human oversight has been delegated to the machine? How do we decompose the engineering effort of self-adaptive systems into manageable subtasks? How can we reuse elements of one adaptive system when constructing another?

The software engineering discipline of self-adaptive systems attempts to answer these questions. It seeks to provide the principles, practices, and tools that will allow engineers to harness the vast potential of adaptation for engineering today's systems.

In doing this, the field of self-adaptive systems has much to draw on from other disciplines: from control theory, techniques for maintaining a system's envelope of behavior within desired ranges; from biology and ecology, the ability of organisms and populations to respond to environmental changes; from immunology, organic mechanisms of self-healing; from software architecture, patterns of structuring systems to enable predictable construction of adaptive systems; from fault tolerance, techniques for detecting and responding to faults; from AI, mechanisms that support autonomy. And many others.

All of this can lead to a rather confusing landscape of concepts and techniques, making it difficult for a software engineer to apply what we know about self-adaptation to the building of software systems. This book by Danny Weyns provides exactly the right introduction for software engineers to navigate this fascinating, complex, and evolving area. It both identifies some foundational principles of the discipline, as well as covering the broad terrain of the field. Through its "waves" approach, it nicely highlights the structure of the field and the influences and perspectives from other disciplines, without losing the fundamental focus on software engineering and applications. Additionally, the waves help to highlight the important research areas that have contributed synergistically to our current understanding of the field and that position us for further advancement.

Taken as a whole, this book provides the first comprehensive treatment of self-adaptive systems targeted at software engineering students, practitioners, and researchers, and provides essential reading for each of these. For someone who is approaching this field for the first time it will provide a broad view of what is now known and practiced. For the experienced professional, it will provide concrete examples and techniques that can be put into practice. For the researcher, it will provide a structured view of the important prior work and of the open challenges facing the field.

February 2020

David Garlan
Professor, School of Computer Science
Carnegie Mellon University

Acknowledgments

This book has been developed in two stages over the past three years. The following colleagues provided me with particularly useful feedback that helped me to improve preliminary versions of this book: Carlo Ghezzi (Politecnico di Milano), Jeff Kramer (Imperial Collage London), Bradley Schmerl (Carnegie Mellon University Pittsburg), Thomas Vogel (Humboldt University Berlin), Gabriel A. Moreno (Carnegie Mellon University), Martina Maggio (Lund University), Antonio Filieri (Imperial Collage London), Marin Litoiu (York University), Vitor E. Silva Souza (Federal University of Espírito Santo), Radu Calinescu (University of York), Jeff Kephart (IBM T. J. Watson Research Center), Betty H.C. Cheng (Michigan State University), Nelly Bencomo (Aston University), Javier Camara Moreno (University of York), John Mylopoulos (University of Toronto), Sebastian Uchitel (University of Buenos Aires), Pooyan Jamshidi Dermani (University of South Carolina), Simos Gerasimou (University of York), Kenji Tei (Waseda University), Dimitri Van Landuyt (Katholieke Universiteit Leuven), Panagiotis (Panos) Patros (University of Waikato), Raffaela Mirandola (Polytechnic University of Milan), and Paola Inverardi (University of L'Aquila). These people have suggested improvements and pointed out mistakes. I thank everyone for providing me with very helpful comments.

I thank the members of the imec-Distrinet research group for their support. I am particularly thankful to my colleague Danny Hughes, his former student Gowri Sankar Ramachandran, and the members of the Network task force for sharing their expertise on the Internet-of-Things. I want to express my sincere appreciation to my colleagues at Linnaeus University, in particular Jesper Andersson, for their continuous support.

I thank M. Usman Iftikhar, Stepan Shevtsov, Federico Quin, Omid Gheibi, Sara Mahdavi Hezavehi, Angelika Musil, Juergen Musil, Nadeem Abbas, and the other students I worked with at KU Leuven and Linnaeus University for their inspiration and collaboration.

I express my sincere appreciation to the monks of the abbeys of West-Vleteren, Westmalle, Tongerlo, and Orval for their hospitality during my stays when working on this book.

Finally, I express my gratitude to Wiley for their support with the publication of this manuscript.

Danny Weyns

Acronyms

24/7	24 hours a day, seven days a week: all the time
A-LTL	Adapt operator-extended Linear Temporal Logic
ActivFORMS	Active FOrmal Models for Self-adaptation
Amazon EC2	Amazon Elastic Compute Cloud
AMOCS-MA	Automated Multi-objective Control of Software with Multiple Actuators
AP	Atomic Propositions
C	Coulomb
CD-ROM	Compact Disk Read Only Memory
CPU	Central Processing Unit
CTL	Computation Tree Logic
dB	deciBel
DCRG	Dynamic Condition Response Graph
DeltaIoT.v2	Advanced version of DeltaIoT
DeltaIoT	IoT application for building security monitoring
DiVA	Dynamic Variability in complex Adaptive systems
DTMC	Discrete-Time Markov Chain
ENTRUST	ENgineering of TRUstworthy Self-adaptive sofTware
EUREMA	ExecUtable RuntimE MegAmodels
F1-score	Score that combines precision and recall to evaluate a classifier
FLAGS	Fuzzy Live Adaptive Goals for Self-adaptive systems
FORMS	FOrmal Reference Model for Self-adaptation
FQL4KE	Fuzzy Q-Learning for Knowledge Evolution
FUSION	FeatUre-oriented Self-adaptatION
GDPR	General Data Protection Regulation
GORE	Goal-Oriented Requirements Engineering
IBM	International Business Machines Corporation
IEEE	Institute of Electrical and Electronics Engineers
IoT	Internet-of-Things
ISO	International Organization for Standardization
KAMI	Keep Alive Models with Implementation
KAOS	Knowledge Acquisition in Automated Specification
LTS	Labeled Transition System
MAPE-K	Monitor-Analyze-Plan-Execute-Knowledge

MARTAS	Models At Runtime And Statistical techniques
mC	mili Coulomb
MDP	Markov Decision Process
MIMO system	Multiple-Input Multiple-Output control system
MIT	Massachusetts Institute of Technology
MJ	Mega Joules
MoRE	Model-Based Reconfiguration Engine
MPC	Model Predictive Control
NATO	North Atlantic Treaty Organization
OSGi	OSGi Alliance, formerly known as the Open Service Gateway Initiative
PCTL	Probabilistic Computation Tree Logic
PI controller	Proportional-Integral controller
PLTS	Probabilistic Labeled Transition System
PRISM	PRobabIlistic Symbolic Model checker
Q-learning	Classic reinforcement learning algorithm
QoS	Quality of Service
QoSMOS	Quality of Service Management and Optimization of Service-based systems
RELAX	Requirements specification language for dynamically adaptive systems
RFID	Radio-Frequency IDentification
RUBiS	Open source auction site prototype modeled after eBay.com
SASO	Stability, Accuracy, Settling time, Overshoot control properties
SAVE	Self-Adaptive Video Encoder
SIMCA	Simplex Control Adaptation
SISO system	Single-Input Single-Output control system
SNR	Signal to Noise Ratio
SSIM	Structural Similarity Index Metric
UML	Unified Modeling Language
Uppaal	Integrated model checking suite for networks of automata
US	United States
UUV	Unmanned Underwater Vehicle
WiFi	Wireless networking technology based on the IEEE 802.11 standards
Z-transform	Transformation of a discrete time to a frequency domain representation
ZNN.com	News service that serves multimedia news content to customers

Introduction

Back in 1968, the North Atlantic Treaty Organization (NATO) organized the first conference on Software Engineering, in Garmisch, Germany. At the time, managers and software engineers perceived a "software crisis," referring to the manageability problems of software projects and software that was not delivering its objectives. One of the key identified causes for the crisis was the growing gap between the rapidly increasing power of computing systems and the ability of programmers to effectively exploit the capabilities of these systems. The crisis was reflected in projects running over-budget and over-time, software of low quality that did not meet requirements, and code that was difficult to maintain. This crisis triggered the development of novel programming paradigms, methods and processes to assure software quality. While today large and complex software projects remain vulnerable to unanticipated problems, the causes that underlay this first software crisis are now relatively well under the control of project managers and software engineers.

About 35 years later, in 2001, IBM released a manifesto that referred to another "looming software crisis," this time caused by the increasing complexity of installing, configuring, tuning, and maintaining computing systems. New emerging computing systems at that time went beyond company boundaries into the Internet, introducing new levels of complexity that could hardly be managed, even by the most skilled system administrators. The complexity resulted from various internal and external factors, causing uncertainties that were difficult to anticipate before deployment. Examples are the scale of the system; inherent distribution of the software system, which may span administrative domains; dynamics in the availability of resources and services; external threats to systems; faults that may be difficult to predict; and changes in user goals during operation. A consensus grew that self-management was the only viable option to tackle the problems that caused this complexity crisis. Self-management refers to computing systems that can adapt autonomously to achieve their goals based on high-level objectives. Such computing systems are usually called *self-adaptive systems*.

From the outset in the early 2000s, there was a common understanding among researchers and engineers that realizing the full potential of self-adaptive systems would take a long-term and worldwide effort across a diversity of fields. Over the past two decades, communities of different fields have put extensive efforts in understanding the foundational principles of self-adaptation as well as devising techniques and methods to engineer self-adaptive systems. This text aims at providing a comprehensive overview of the field of self-adaptation by consolidating key knowledge obtained from these efforts.

Introducing self-adaptive systems is challenging given the diversity of research topics, engineering methods, and application domains that are part of this field. To tackle this challenge, this text is based on six pillars.

First, we lay a foundation for *what* constitutes a self-adaptive system by introducing two generally acknowledged, but complementary basic principles. These two principles enable us to characterize self-adaptive systems and distinguish them from other related types of systems. From the basic principles, a conceptual model of a self-adaptive system is derived, which offers a basic vocabulary that we use throughout the text.

Second, the core of the text, which focuses on *how* self-adaptive systems are engineered, is partitioned into convenient chunks driven by research and engineering efforts over time. In particular, the text approaches the engineering of self-adaptive systems in seven waves. These waves put complementary aspects of engineering self-adaptive systems in focus that synergistically have contributed to the current body of knowledge in the field. Each wave highlights a trend of interest in the research community. Some of the earlier waves have stabilized now and resulted in common knowledge in the community. Other more recent waves are still very active and the subject of debate; the knowledge of these waves has not been fully consolidated yet.

Third, throughout the text we use a well-thought-out set of applications to illustrate the material with concrete examples. We use a simple service-based application to illustrate the basic principles and the conceptional model of self-adaptive systems. Before the core part of the text that zooms in on the seven waves of research on engineering self-adaptive systems, we introduce a practical Internet-of-Things application that we use as the main case to illustrate the characteristics of the different waves. In addition, we use a variety of cases from different contemporary domains to illustrate the material, including a client-server system, a mobile service, a geo-localization service, unmanned vehicles, video compression, different Web applications, and a Cloud system.

Fourth, each core chapter of the book starts with a list of learning outcomes at different orders of thinking (from understanding to synthesis) and concludes with a series of exercises. The exercises are defined at four different levels of complexity, characterized by four letters that refer to the expected average time required for solving the exercises. Level H requires a basic understanding of the material of the chapter; the exercises should be solvable in a number of person-hours. Level D requires in depth understanding of the material of the chapter; these exercises should be solvable within person-days. Level W requires the study of some additional material beyond the material in the chapter; these exercises should be solvable within person-weeks. Finally, level M requires the development of novel solutions based on the material provided in the corresponding chapter; these exercises require an effort of person-months. The final chapter discusses the maturity of the field and outlines open challenges for research in self-adaptation, which can serve as further inspiration for future research endeavors, for instance as a start point for PhD projects.

Fifth, each chapter concludes with bibliographic notes. These notes point to foundational research papers of the different parts of the chapter. In addition, the notes highlight some characteristic work and provide pointers to background material. The material referred to in the bibliographic notes is advised for further reading.

Sixth, supplementary material is freely available for readers, students, and teachers at the book website: **https://introsas.cs.kuleuven.be/**. The supplementary material includes slides for educational purposes, selected example solutions of exercises, models and code that can be used for the exercises, and complementary material that elaborates on specific material from the book.

As such, this manuscript provides a starting point for students, researchers, and engineers that want to familiarize themselves with the field of self-adaptation. The text aims to offer a solid basis for those who are interested in self-adaptation to obtain the required skill set to understand the fundamental principles and engineering methods of self-adaptive systems.

The principles of self-adaptation have their roots in software architecture, model-based engineering, formal specification languages, and principles of control theory and machine learning. It is expected that readers are familiar with the basics of these topics when starting with our book, although some basic aspects are introduced in the respective chapters.

1

Basic Principles of Self-Adaptation and Conceptual Model

Modern software-intensive systems[1] are expected to operate under uncertain conditions, without interruption. Possible causes of uncertainties include changes in the operational environment, dynamics in the availability of resources, and variations of user goals. Traditionally, it is the task of system operators to deal with such uncertainties. However, such management tasks can be complex, error-prone, and expensive. The aim of self-adaptation is to let the system collect additional data about the uncertainties during operation in order to manage itself based on high-level goals. The system uses the additional data to resolve uncertainties and based on its goals re-configures or adjusts itself to satisfy the changing conditions.

Consider as an example a simple service-based health assistance system as shown in Figure 1.1. The system takes samples of vital parameters of patients; it also enables patients to invoke a panic button in case of an emergency. The parameters are analyzed by a medical service that may invoke additional services to take actions when needed; for instance, a drug service may need to notify a local pharmacy to deliver new medication to a patient. Each service type can be realized by one of multiple service instances provided by third-party service providers. These service instances are characterized by different quality properties, such as failure rate and cost. Typical examples of uncertainties in this system are the patterns that particular paths in the workflow are invoked by, which are based on the health conditions of the users and their behavior. Other uncertainties are the available service instances, their actual failure rates and the costs to use them. These parameters may change over time, for instance due to the changing workloads or unexpected network failures.

Anticipating such uncertainties during system development, or letting system operators deal with them during operation, is often difficult, inefficient, or too costly. Moreover, since many software-intensive systems today need to be operational 24/7, the uncertainties necessarily need to be resolved at runtime when the missing knowledge becomes available. Self-adaptation is about how a system can mitigate such uncertainties autonomously or with minimum human intervention.

The basic idea of self-adaptation is to let the system collect new data (that was missing before deployment) during operation when it becomes available. The system uses the

---

1 A software-intensive system is any system where software dominates to a large extent the design, construction, deployment, operation, and evolution of the system. Some examples include mobile embedded systems, unmanned vehicles, web service applications, wireless ad-hoc systems, telecommunications, and Cloud systems.

An Introduction to Self-Adaptive Systems: A Contemporary Software Engineering Perspective,
First Edition. Danny Weyns.
© 2021 John Wiley & Sons Ltd. Published 2021 by John Wiley & Sons Ltd.

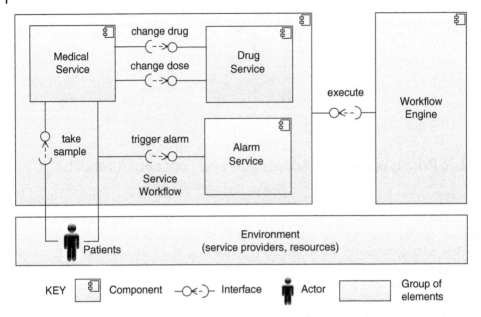

Figure 1.1 Architecture of a simple service-based health assistance system

additional data to resolve uncertainties, to reason about itself, and based on its goals to reconfigure or adjust itself to maintain its quality requirements or, if necessary, to degrade gracefully.

In this chapter, we explain *what* a self-adaptive system is. We define two basic principles that determine the essential characteristics of self-adaptation. These principles allow us to define the boundaries of what we mean by a self-adaptive system in this book, and to contrast self-adaptation with other approaches that deal with changing conditions during operation. From the two principles, we derive a conceptual model of a self-adaptive system that defines the basic elements of such a system. The conceptual model provides a basic vocabulary for the remainder of this book.

LEARNING OUTCOMES

- To explain the basic principles of self-adaptation.
- To understand how self-adaptation relates to other adaptation approaches.
- To describe the conceptual model of a self-adaptive system.
- To explain and illustrate the basic concepts of a self-adaptive system.
- To apply the conceptual model to a concrete self-adaptive application.

1.1 Principles of Self-Adaptation

There is no general agreement on a definition of the notion of *self-adaptation*. However, there are two common interpretations of what constitutes a self-adaptive system.

The first interpretation considers a self-adaptive system as a system that is able to adjust its behavior in response to the perception of changes in the environment and the system itself. The *self* prefix indicates that the system decides autonomously (i.e. without or with minimal human intervention) how to adapt to accommodate changes in its context and environment. Furthermore, a prevalent aspect of this first interpretation is the presence of uncertainty in the environment or the domain in which the software is deployed. To deal with these uncertainties, the self-adaptive system performs tasks that are traditionally done by operators. Hence, the first interpretation takes the stance of the external observer and looks at a self-adaptive system as a black box. Self-adaptation is considered as an observable property of a system that enables it to handle changes in external conditions, availability of resources, workloads, demands, and failures and threats.

The second interpretation contrasts traditional "internal" mechanisms that enable a system to deal with unexpected or unwanted events, such as exceptions in programming languages and fault-tolerant protocols, with "external" mechanisms that are realized by means of a closed feedback loop that monitors and adapts the system behavior at runtime. This interpretation emphasizes a "disciplined split" between two distinct parts of a self-adaptive system: one part that deals with the domain concerns and another part that deals with the adaptation concerns. Domain concerns relate to the goals of the users for which the system is built; adaptation concerns relate to the system itself, i.e. the way the system realizes the user goals under changing conditions. The second interpretation takes the stance of the engineer of the system and looks at self-adaptation from the point of view how the system is conceived.

Hence, we introduce *two complementary basic principles* that determine what a self-adaptive system is:

---

1. **External principle:** A self-adaptive system is a system that can handle changes and uncertainties in its environment, the system itself, and its goals autonomously (i.e. without or with minimal required human intervention).
2. **Internal principle:** A self-adaptive system comprises two distinct parts: the first part interacts with the environment and is responsible for the domain concerns – i.e. the concerns of users for which the system is built; the second part consists of a feedback loop that interacts with the first part (and monitors its environment) and is responsible for the adaptation concerns – i.e. concerns about the domain concerns.

Let us illustrate how the two principles of self-adaptation apply to the service-based health assistance system. Self-adaptation would enable the system to deal with dynamics in the types of services that are invoked by the system as well as variations in the failure rates and costs of particular service instances. Such uncertainties may be hard to anticipate before the system is deployed (external principle). To that end, the service-based system could be enhanced with a feedback loop. This feedback loop tracks the paths of services that are invoked in the workflow, as well as the failure rates of service instances and the costs of invoking service instances that are provided by the service providers. Taking this data into account, the feedback loop adapts the selection of service instances by the workflow engine such that a set of adaptation concerns is achieved. For instance, services are selected that keep the average failure rate below a required threshold, while the cost of using the health assistance system is minimized (internal principle).

1.2 Other Adaptation Approaches

The ability of a software-intensive system to adapt at runtime in order to achieve its goals under changing conditions is not the exclusivity of self-adaptation, but can be realized in other ways.

The field of autonomous systems has a long tradition of studying systems that can change their behavior during operation in response to events that may not have been anticipated fully. A central idea of autonomous systems is to mimic human (or animal) behavior, which has been a source of inspiration for a very long time. The area of cybernetics founded by Norbert Wiener at MIT in the mid twentieth century led to the development of various types of machines that exposed seemingly "intelligent" behavior similar to biological systems. Wiener's work contributed to the foundations of various fields, including feedback control, automation, and robotics. The interest in autonomous systems has expanded significantly in recent years, with high-profile application domains such as autonomous vehicles. While these applications have extreme potential, their successes so far have also been accompanied by some dramatic failures, such as the accidents caused by first generation autonomous cars. The consequences of such failures demonstrate the real technical difficulties associated with realizing truly autonomous systems.

An important sub-field of autonomous systems is multi-agent systems, which studies the coordination of autonomous behavior of agents to solve problems that go beyond the capabilities of single agents. This study involves architectures of autonomous agents, communication and coordination mechanisms, and supporting infrastructure. An important aspect is the representation of knowledge and its use to coordinate autonomous behavior of agents. Self-organizing systems emphasize decentralized control. In a self-organizing system, simple reactive agents apply local rules to adapt their interactions with other agents in response to changing conditions in order to cooperatively realize the system goals. In such systems, the global macroscopic behavior emerges from the local interactions of the agents. However, emergent behavior can also appear as an unwanted side effect, for example in the form of oscillations. Designing decentralized systems that expose the required global behavior while avoiding unwanted emergent phenomena remains a major challenge.

Context-awareness is another traditional field that is related to self-adaptation. Context-awareness puts the emphasis on handling relevant elements in the physical environment as first-class citizens in system design and operation. Context-aware computing systems are concerned with the acquisition of context (e.g. through sensors to perceive a situation), the representation and understanding of context, and the steering of behavior based on the recognized context (e.g. triggering actions based on the actual context). Context-aware systems typically have a layered architecture, where a context manager or dedicated middleware is responsible for sensing and dealing with context changes. Self-aware computing systems contrast with context-aware computing systems in the sense that these systems capture and learn knowledge not only about the environment but also about themselves. This knowledge is encoded in the form of runtime models, which a self-aware system uses to reason at runtime, enabling it to act in accordance with higher-level goals.

1.3 Scope of Self-Adaptation

Autonomous systems, multi-agent systems, self-organizing systems, and context-aware systems are families of systems that apply classical approaches to deal with change at runtime. However, these approaches do not align with the combined basic principles of self-adaptation. In particular, none of these approaches comply with the second principle, which makes an explicit distinction between a part of the system that handles domain concerns and a part that handles adaptation concerns. However, the second principle of self-adaptation can be applied to each of these approaches – i.e. these systems can be enhanced with a feedback loop that deals with a set of adaptation concerns. This book is concerned with self-adaptation as a property of a computing system that is compliant with the two basic principles of self-adaptation.

Furthermore, self-adaptation can be applied at different levels of the software stack of computing systems, from the underlying resources and low-level computing infrastructure to middleware services and application software. The challenges of self-adaptation at these different levels are different. For instance, the space of adaptation options of higher-level software entities is often multi-dimensional, and software qualities and adaptation goals usually have a complex interplay. These characteristics are less applicable to the adaptation of lower-level resources, where there is often a more straightforward relation between adaptation actions and software qualities. In this book, we consider self-adaptation applied at different levels of the software stack of computing systems, from virtualized resources up to application software.

1.4 Conceptual Model of a Self-Adaptive System

Starting from the two basic principles of self-adaptation, we define a conceptual model for self-adaptive systems that describes the basic elements of such systems and the relationship between them. The basic elements are intentionally kept abstract and general, but they are compliant with the basic principles of self-adaptation. The conceptual model introduces a basic vocabulary for the field of self-adaptation that we will use throughout this book. Figure 1.2 shows the conceptual model of a self-adaptive system.

The conceptual model comprises *four basic elements:* environment, managed system, feedback loop, and adaptation goals. The feedback loop together with the adaptation goals form the managing system. We discuss the elements one by one and illustrate them for the service-based health assistance application.

1.4.1 Environment

The environment refers to the part of the external world with which a self-adaptive system interacts and in which the effects of the system will be observed and evaluated. The environment can include users as well as physical and virtual elements. The distinction between

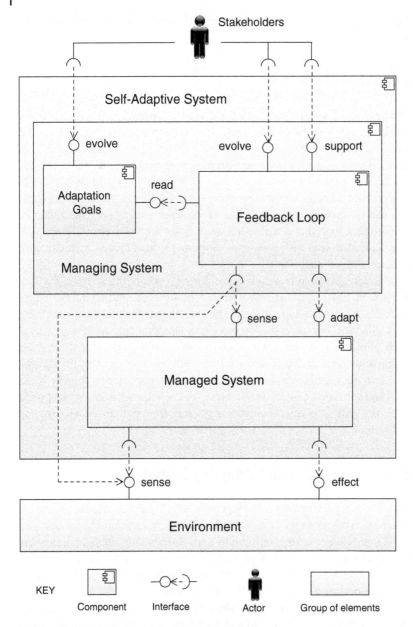

Figure 1.2 Conceptual model of a self-adaptive system

the environment and the self-adaptive system is made based on the extent of control. The environment can be sensed and effected through sensors and effectors, respectively. However, as the environment is not under the control of the software engineer of the system, there may be uncertainty in terms of what is sensed by the sensors or what the outcomes will be of the effectors.

Applied to the service-based health assistance system example, the environment includes the patients that make use of the system; the application devices with the sensors that measure vital parameters of patients and the panic buttons; the service providers with the services instances they offer; and the network connections used in the system, which may all affect the quality properties of the system.

1.4.2 Managed System

The managed system comprises the application software that realizes the functions of the system to its users. Hence, the concerns of the managed system are concerns over the domain, i.e. the environment of the system. Different terminology has been used to refer to the managed system, such as managed element, system layer, core function, base-level system, and controllable plant. In this book, we systematically use the term *managed system*. To realize its functions to the users, the managed system senses and effects the environment. To support adaptations, the managed system needs to be equipped with sensors to enable monitoring and effectors (also called actuators) to execute adaptation actions. Safely executing adaptations requires that actions applied to the managed systems do not interfere with the regular system activity. In general, they may affect ongoing activities of the system – for instance, scaling a Cloud system might require bringing down a container and restarting it.

A classic approach to realizing safe adaptations is to apply adaptation actions only when a system (or the parts that are subject to adaptation) is in a *quiescent state*. A quiescent state is a state where no activity is going on in the managed system or the parts of it that are subject to adaptation so that the system can be safely updated. Support for quiescence requires an infrastructure to deal with messages that are invoked during adaptations; this infrastructure also needs to handle the state of the adapted system or the relevant parts of it to ensure its consistency before and after adaptation. Handling such messages and ensuring consistency of state during adaptations are in general difficult problems. However, numerous infrastructures have been developed to support safe adaptations for particular settings. A well-known example is the OSGi (Open Service Gateway Initiative) Java framework, which supports installing, starting, stopping, and updating arbitrary components (bundles in OSGi terminology) dynamically.

The managed system of the service-based health assistance system consists of a service workflow that realizes the system functions. In particular, a medical service receives messages from patients with values of their vital parameters. The service analyzes the data and either invokes a drug service to notify a local pharmacy to deliver new medication to the patient or change the dose of medication, or it invokes an alarm service in case of an emergency to notify medical staff to visit the patient. The alarm service can also be invoked directly by a patient via a panic button. To support adaptation, the workflow infrastructure offers sensors to track the relevant aspects of the system and the characteristics of service instances (failure rate and cost). The infrastructure allows the selection and use of concrete instances of the different types of services that are required by the system. Finally, the workflow infrastructure needs to provide support to change service instances in a consistent manner by ensuring that a service is only removed and replaced when it is no longer involved in any ongoing service invocation of the health assistance system.

1.4.3 Adaptation Goals

Adaptation goals represent concerns of the managing system over the managed system; adaptation goals relate to quality properties of the managed system. In general, four principal types of high-level adaptation goals can be distinguished: self-configuration (i.e. systems that configure themselves automatically), self-optimization (systems that continually seek ways to improve their performance or reduce their cost), self-healing (systems that detect, diagnose, and repair problems resulting from bugs or failures), and self-protection (systems that defend themselves from malicious attacks or cascading failures).

Since the system uses the adaptation goals to reason about itself during operation, the goals need to be represented in a machine-readable format. Adaptation goals are often expressed in terms of the uncertainty they have to deal with. Example approaches are the specification of quality of service goals using probabilistic temporal logics that allow for probabilistic quantification of properties, the specification of fuzzy goals whose satisfaction is represented through fuzzy constraints, and a declarative specification of goals (in contrast to enumeration) allowing the introduction of flexibility in the specification of goals. Adaptation goals can be subject to change themselves, which is represented in Figure 1.2 by means of the *evolve* interface. Adding new goals or removing goals during operation will require updates of the managing system, and often also require updates of probes and effectors.

In the health assistance application, the system dynamically selects service instances under changing conditions to keep the failure rate over a given period below a required threshold (self-healing goal), while the cost is minimized (optimization goal). Stakeholders may change the threshold value for the failure rate during operation, which may require just a simple update of the corresponding threshold value. On the other hand, adding a new adaptation goal, for instance to keep the average response time of invocations of the assistance service below a required threshold, would be more invasive and would require an evolution of the adaptation goals and the managing system.

1.4.4 Feedback Loop

The adaptation of the managed system is realized by the managing system. Different terms are used in the literature for the concept of managing system, such as autonomic manager, adaptation engine, reflective system, and controller. Conceptually, the managing system realizes a feedback loop that manages the managed system. The feedback loop comprises the adaptation logic that deals with one or more adaptation goals. To realize the adaptation goals, the feedback loop monitors the environment and the managed system and adapts the latter when necessary to realize the adaptation goals. With a reactive policy, the feedback loop responds to a violation of the adaptation goals by adapting the managed system to a new configuration that complies with the adaptation goals. With a proactive policy, the feedback loop tracks the behavior of the managed system and adapts the system to anticipate a possible violation of the adaptation goals.

An important requirement of a managing system is ensuring that fail-safe operating modes are always satisfied. When such an operating mode is detected, the managing system can switch to a fall-back or degraded mode during operation. An example of an operating mode that may require the managing system to switch to a fail-safe configuration is the inability to find a new configuration to adapt the managed system to that achieves the adaptation goals within the time window that is available to make an adaptation decision. Note that instead of falling back to a fail-safe configuration in the event that the goals cannot be achieved, the managing system may also offer a stakeholder the possibility to decide on the action to take.

The managing system may consist of a single level that conceptually consists of one feed-back loop with a set of adaptation goals, as shown in Figure 1.2. However, the managing system may also have a layered structure, where each layer conceptually consists of a feed-back loop with its own goals. In this case, each layer manages the layer beneath – i.e. layer n manages layer n-1, and layer 1 manages the managed system. In practice, most self-adaptive systems have a managing system that consists of just one layer. In systems where additional layers are applied, the number of additional layers is usually limited to one or two. For instance, a managing system may have two layers: the bottom layer may react quickly to changes and adapts the managed system when needed, while the top layer may reason over long term strategies and adapt the underlying layer accordingly.

The managing system can operate completely automatically without intervention of stakeholders, or stakeholders may be involved in support for certain functions realized by the feedback loop; this is shown in Figure 1.2 by means of the generic *support* interface. We already gave an example above where a stakeholder could support the system with handling a fail-safe situation. Another example is a managing system that detects a possible threat to the system. Before activating a possible reconfiguration to mitigate the threat, the managing system may check with a stakeholder whether the adaptation should be applied or not.

The managing system can be subject to change itself, which is represented in Figure 1.2 with the *evolve* interface. On-the-fly changes of the managing systems are important for two main reasons: (i) to update a feedback loop to resolve a problem or a bug (e.g. add or replace some functionality), and (ii) to support changing adaptation goals, i.e. change or remove an existing goal or add a new goal. The need for evolving the feedback loop model is triggered by stakeholders either based on observations obtained from the executing system or because stakeholders want to change the adaptation goals.

The managing system of the service-based health assistance system comprises a feedback loop that is added to the service workflow. The task of the feedback loop is to ensure that the adaptation goals are realized. To that end, the feedback loop monitors the system behavior and the quality properties of service instances, and tracks that the system is not violating the adaptation goals. For a reactive policy, the feedback loop will select alternative service instances that ensure the adaptation goals are met in the event that goal violations are detected. If no configuration can be found that complies with the adaptation goals within a given time (fail-safe operating mode), the managing system may involve a stakeholder to decide on the adaptation action to take. The feedback loop that adapts the service instances to ensure that the adaptation goals are realized may be extended with an extra level that adapts the underlying method that makes the adaptation decisions. For instance, this extra level may track the quality properties of service instances over time and identify patterns. The second layer can then use this knowledge to instruct the underlying feedback loop to give preference to selecting particular service instances or to avoid the selection of certain instances. For instance, services that expose a high level of failures during particular periods of the day may temporarily be excluded from selection to avoid harming the trustworthiness of the system. As we explained above, when a new adaptation goal is added to the system, in order to keep the average latency of invocations of the assistance service below a required threshold, the managing system will need to be updated. For instance, the managing system will need to be updated such that it can make adaptation decisions based on three adaptation goals instead of two.

1.4.5 Conceptual Model Applied

Figure 1.3 summarizes how the the conceptual model maps to the self-adaptive service-based health assistance system. The operator in this particular instance is responsible for supporting the self-adaptive system with handling fail-safe conditions (through the support interface). In this example, we do not consider the evolution of adaptation goals and the managing system.

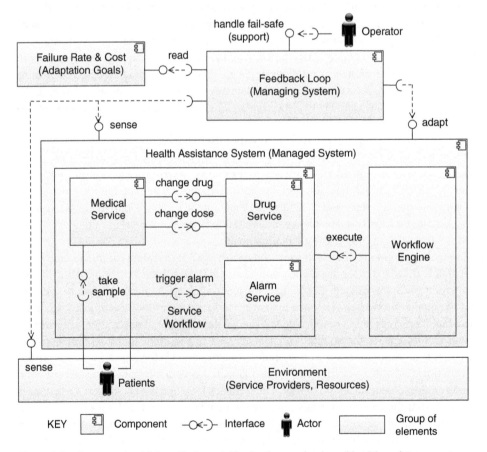

Figure 1.3 Conceptual model applied to a self-adaptive service-based health assistance system

1.5   A Note on Model Abstractions

It is important to note that the conceptual model for self-adaptive systems abstracts away from distribution – i.e. the deployment of the software to hardware that is connected via a network. Whereas a distributed self-adaptive system consists of multiple software components that are deployed on multiple nodes connected via some network, from a conceptual point of view such system can be represented as one managed system (that deals with the domain concerns) and one managing system (that deals with adaptation concerns of the managed system). The conceptual model also abstracts away from how adaptation decisions in a self-adaptive system are made and potentially coordinated among different components. In particular, the conceptual model is invariant to self-adaptive systems where the adaptation functions are made by a single centralized entity or by multiple coordinating entities in a decentralized way. In a concrete setting, the composition of the components of a self-adaptive system, the concrete deployment of these components to hardware elements, and the degree of decentralization of the decision making of adaptation will have a deep impact on how such self-adaptive systems are engineered.

1.6   Summary

Dealing with uncertainties in the operating conditions of a software-intensive system that are difficult to predict is an important challenge for software engineers. Self-adaptation is about how a system can mitigate such uncertainties.

There are two common interpretations of what constitutes a self-adaptive system. The first interpretation considers a self-adaptive system as a system that is able to adjust its behavior in response to changes in the environment or the system itself. The second interpretation contrasts traditional internal mechanisms that enable a system to deal with unexpected or unwanted events with external mechanisms that are realized by means of feedback loops.

These interpretations lead to two complementary basic principles that determine what is a self-adaptive system. The external principle states that a self-adaptive system can handle change and uncertainties autonomously (or with minimal human intervention). The internal principle states that a self-adaptive system consists of two distinct parts: one part that interacts with the environment and deals with the domain concerns and a second part that interacts with the first part and deals with the adaptation concerns.

Other traditional approaches to deal with change at runtime include autonomous systems, multi-agent systems, self-organizing systems, and context-aware systems. These approaches differ from self-adaptation, in particular with respect to the second basic principle. However, the second principle can be applied to these approaches through adding a managing system realizing self-adaptation.

Conceptually, a self-adaptive system consists of four basic elements: environment, managed system, adaptation goals, and feedback loop. The environment is external to the system; it defines the domain concerns and is not under control of the software engineer. The managed system comprises the application software that realizes the domain concerns for the users. To support adaptation, the managed system needs to provide probes and effectors and support safe adaptations. The adaptation goals represent concerns over the managed system, which refer to qualities of the system. The feedback loop realizes the adaptation goals by monitoring and adapting the managed system. The feedback loop with the adaptation goals form the managing system. The managing system can be subject to on-the-fly evolution, either to update some functionality of the adaptation logic or to change the adaptation goals.

1.7 Exercises

1.1 Conceptual model pipe and filter system: level H

Consider a pipe and filter system that has to perform a series of tasks for a user. Different instances of the filters are offered by third parties. These filter instances provide different quality of service in terms of processing time and service cost that may change over time. Explain how you would make this a self-adaptive system that ensures that the average throughput of tasks remains under a given threshold while the cost is minimized. Draw the conceptual model that shows your solution to this adaptation problem.

1.2 Conceptual model Znn.com news service: level H

Setting. Consider Znn.com, a news service that serves multimedia news content to customers. Architecturally, Znn.com is set up as a Web-based client-server system that serves clients from a pool of servers. Customers of Znn.com expect a reasonable response time, while the system owner wants to keep the cost of the server pool within a certain operating budget. In normal operating circumstances, the appropriate trade-offs can be made at design-time. However, from time to time, due to highly popular events, Znn.com experiences spikes in news requests that are not within the originally designed parameters. This means that the clients will not receive content in a timely manner. To the clients, the site will appear to be down, so they may not use the service anymore, resulting in lost revenue. The challenge for self-adaptation is to enable the system to still provide content at peak times. There are several ways to deal with this, such as serving reduced content, increasing the number of servers serving content, and choosing to prioritize serving paying customers.

Task. Enhance Znn.com with self-adaptation to deal with the challenge of the news service. Identify the basic concepts of the self-adaptive system (environment, managed system, feedback loop, adaptation goals) and describe the responsibilities of each element. Draw the conceptual model that shows your solution to this adaptation problem.

Additional material. See the Znn artifact website [53].

1.3 Conceptual model video encoder: level H

Setting. Consider a video encoder that takes a stream of video frames (for instance from an mp4 video) and compresses the frames such that the video stream fits a given communication channel. While compressing frames, the encoder should maintain a required quality of the manipulated frames compared to the original frames, which is expressed as a similarity index. To achieve these conflicting goals, the encoder can change three parameters for each frame: the quality of the encoding and the setting of a sharpening filter and the setting of a noise reduction filter that are both applied to the image. The quality parameter that relates to a compression factor for the image has a value between 1 and 100, where 100 preserves all frame details and 1 produces the highest compression. However, the relationship between quality and size depends on the frame content, which is difficult to predict upfront. The sharpening filter and the noise reduction filter modify certain pixels of the imagine, for instance to remove elements that appear after compressing the original frame. The sharpening filter has a parameter with a value that ranges between 0 and 5, where 0 indicates no sharpening and 5 maximum sharpening. The noise reduction filter has a parameter that specifies the size of the applied noise reduction filter, which also varies between 0 and 5.

Task. Enhance the video encoder with self-adaptation capabilities to deal with the conflicting goals of compressing frames and ensuring a required level of quality. Identify the basic concepts of the self-adaptive system (environment, managed system, feedback loop, adaptation goals) and describe the responsibilities of each element. Draw the conceptual model that shows your solution to this adaptation problem.

Additional material. See the Self-Adaptive Video Encoder artifact website [136].

1.4 Implementation feedback loop Tele-Assistance System: level D

Setting. TAS, short for Tele-Assistance System, is a Java-based artifact that supports research and experimentation on self-adaptation. TAS simulates a health assistance service for elderly and chronically sick people, similar to the health assistance service used in this chapter. TAS uses a combination of sensors embedded in a wearable device and remote third-party services from medical analysis, pharmacy and emergency service providers. The TAS workflow periodically takes measurements of the vital parameters of a patient and employs a medical service for their analysis. The result of an analysis may trigger the invocation of a pharmacy service to deliver new medication to the patient or to change their dose of medication, or, in a critical situation, the invocation of an alarm service that will send a medical assistance team to the patient. The same alarm service can be invoked directly by the patient by using a panic button on the wearable device. In practice, the TAS service will be subject to a variety of uncertainties: services may fail, service response times may vary, or new services may become available. Different types of adaptations can be applied to deal with these uncertainties, such as switching to equivalent services, simultaneously invoking several services for equivalent operations, or changing the workflow architecture.

Task. Download the source code of TAS. Read the developers guide that is part of the artifact distribution, and prepare Eclipse to work with the artifact. Execute the TAS artifact and get familiar with it. Now design a feedback loop that deals with service failures. The first adaptation goal is a threshold goal that requires that the average number of service failures should not exceed 10% of the invocations over 100 service invocations. The second adaptation goal is to minimize the cost for service invocations over 100 service invocations. Implement your design and test it. Evaluate your solution and assess.

Additional material. For the TAS artifact, see [201]. The latest version of TAS can be downloaded from the TAS website [212]. For background information about TAS, see [200].

1.8   Bibliographic Notes

The external principle of self-adaptation is grounded in the description of what constitutes a self-adaptive system provided in a roadmap paper on engineering self-adaptive system [50]. Y. Brun et al. complemented this description and motivated the "self" prefix indicating that the system decides autonomously [35]. The internal principle of self-adaptation is grounded in the pioneering work of P. Oreizy et al. that stressed the need for a systematic approach to deal with software modification at runtime (as opposed to ad-hoc "patches") [150]. In their seminal work on Rainbow, D. Garlan et al. contrasted internal mechanisms to adapt a system (for instance using exceptions) with external mechanisms that enhance a system with an external feedback loop that is responsible for handling adaptation [81].

Back in 1948, N. Wiener published a book that coined the term "cybernetics" to refer to self-regulating mechanisms. This work laid the theoretical foundation for several fields in autonomous systems. M. Wooldridge provided a comprehensive and readable introduction to the theory and practice of the field of multi-agent systems [215]. F. Heylighen reviewed the most important concepts and principles of self-organization [97]. Based on these principles, V. Dyke Parunak et al. demonstrated how digital pheromones enable robust coordination between unmanned vehicles [190]. T. De Wolf and T. Holvoet contrast self-organization with emergent behavior [60].

B. Schilit et al. defined the notion of context-aware computing and described different categories of context-aware applications [172]. In the context of autonomic systems, Hinchey and Sterritt referred to self-awareness as the capability of a system to be aware of its states and behaviors [98]. M. Parashar and S. Hariri referred to self-awareness as the ability of a system to be aware of its operational environment [153]. P. Gandodhar et al. reported the results of a survey on context-awarenss [79], and C. Perera et al. surveyed context-aware computing in the area of the Internet-of-Things [154]. S. Kounev et al. defined self-aware computing systems and outlined a taxonomy for these types of systems [119].

Several authors have provided arguments for why engineering self-adaptation at different levels of the technology stack poses different challenges. Among these are the growing complexity of the adaptation space from lower-level resources up to higher-level software [5, 36], and the increasingly complex interplay between system qualities on the one hand and adaptation options at higher levels of the software stack on the other hand [72].

M. Jackson contrasted the notion of environment, which is not under the control of a designer, and the system, which is controllable [106]. J. Kramer and J. Magee introduced the notion of quiescence [120]. A quiescent state of a software element is a state where no activity is going on in the element so that it can be safely updated. Such a state may be reached spontaneously or it may need to be enforced. J. Zhang and B. Cheng created the A-LTL specification language to specify the semantics of adaptive programs [218], underpinning safe adaptations. The OSGi framework [2] offers a modular service platform for Java that implements a dynamic component model that allows components (so called bundles) to be installed, started, stopped, updated, and uninstalled without requiring a reboot.

J. Kephart and D. Chess identified the primary types of higher-level adaptation goals [112]: self-configuration, self-optimization, self-healing, and self-protection.

M. Salehie and L. Tahvildari referred to self-adaptive software as software that embodies a closed-loop mechanism in the form of an adaptation loop [170]. Similarly, Dobson et al. referred to an autonomic control loop, which includes processes to collect and analyze data, and decide and act upon the system [65]. Y. Brun et al. argued for making feedback loops first-class entities in the design and operation of self-adaptive systems [35].

J. Camara et al. elaborated on involving humans in the feedback loop to support different self-adaptation functions, including the decision-making process [44]. Weyns et al. presented a set of architectural patterns for decentralizing control in self-adaptive systems [209].

The service-based health assistance system used in this book is based on the Tele-Assistance System (TAS) exemplar [200]. TAS offers a prototypical application that can be used to evaluate and compare new methods, techniques, and tools for research on self-adaptation in the domain of service-based systems. The service-based health assistance system was originally introduced in [15].

2

Engineering Self-Adaptive Systems: A Short Tour in Seven Waves

Handling change is an increasingly important challenge for software engineers. Our focus is on changes caused by uncertainties in the operating conditions of systems that are hard to anticipate before deployment or for operators to manage during operation. Such uncertainties may jeopardize the realization of the system requirements. Self-adaptation aims at mitigating these uncertainties without human intervention or with minimal intervention if interaction with stakeholders is required. A characteristic example of a self-adaptive system is a Cloud platform that monitors the varying load of client applications and automatically adjusts capacity to maintain steady performance at the lowest possible cost.

In Chapter 1, we focused on *what* self-adaptation is. We explained the two basic principles of self-adaptation and outlined a conceptual model that describes the basic elements of a self-adaptive system that is compliant with the basic principles. We direct our focus now to *how* self-adaptive systems are engineered.

In general, engineering software-intensive systems that can handle uncertainty is complex as uncertainty can affect many aspects of a system, including its context, goals, models, and functional and quality properties. From a software engineering process point of view, in a self-adaptive system, change activities are shifted from development time to runtime, and the responsibilities for these activities are shifted from software engineers or system administrators to the system itself. Therefore the traditional boundary between development time and runtime becomes dynamic and fades away. Different software activities that are traditionally performed by software engineers during system design and development are now partially or completely pushed into the runtime and need to be managed by the system itself, either fully autonomously or, if required, supported by stakeholders. This implies the need for a reconceptualization of different activities of the software engineering process.

In the following chapters, we provide a comprehensive and in-depth introduction to the engineering of self-adaptive systems. Instead of presenting distinct approaches for engineering self-adaptive systems that have been studied and applied over time, we take a different stance on the field and put different aspects of engineering self-adaptive systems into focus. In particular, our introduction to the engineering of self-adaptive systems is structured into seven waves that have emerged over time. These waves put *complementary aspects of research on engineering self-adaptive systems* into focus that synergistically have contributed to the current body of knowledge in the field. Each wave highlights a trend of interest in the community of self-adaptive systems research. More recent waves were triggered by insights derived from earlier waves. Some of the earlier waves have stabilized now and resulted in common knowledge used in the community. Other, usually more recent, waves are still very active and the subject of debate; the knowledge of these waves has not been consolidated yet.

In this chapter, we make a short tour through the seven waves before we explore the different waves in detail in the following chapters.

LEARNING OUTCOMES

- To provide a short tour through the seven waves of engineering self-adaptive systems.
- To summarize the focus of each of the seven waves.
- To explain the state of the art before each wave and the contributions enabled by each wave.

2.1 Overview of the Waves

Figure 2.1 shows a schematic overview of the seven waves of research in the field of self-adaptive systems that have emerged over time. Each wave is represented by an oval, and they are numbered chronologically, loosely based on the time they emerged. The arrows indicate how insights from earlier waves have initiated or triggered new waves.

The first wave, *Automating Tasks*, is concerned with delegating complex and error-prone management tasks from human operators to the system. The automation of difficult management tasks is a principal driver for self-adaptation. To that end, the system is extended with an external manager (i.e. a managing system) that tracks the system and automatically adapts it to changing conditions based on high-level objectives provided by administrators. A central aspect of this wave is the decomposition of the external manager into the basic functions that it needs to realize.

The second wave, *Architecture-based Adaptation*, which is triggered by the need for a systematic engineering approach (from the first wave), is concerned with applying the architectural principles of abstraction and separation of concerns to identify the foundations of engineering self-adaptive systems. On the one hand, architecture offers abstractions to define self-adaptive systems. Central to this definition is the separation between change management (dealing with changing operating conditions) and goal management (dealing with changing goals). On the other hand, architecture offers modeling abstractions that enable the system to reason about change during operation. Central to this reasoning is an architectural model that offers an appropriate perspective on the system and its environment enabling the system to make effective adaptation decisions.

The third wave, *Runtime Models*, is triggered by the problem of how to manage the complexity of realizing concrete designs of self-adaptive systems (from the second wave). This wave enhances the exploitation of first-class runtime models that represent the different key elements of a self-adaptive system that are needed to reason about adaptation during operation. The third wave extends the notion of development models, which are used in model-driven development of software, to the runtime. These runtime models provide a view on the different aspects of a software system in operation that can be used by the system to make adaptation decisions at runtime.

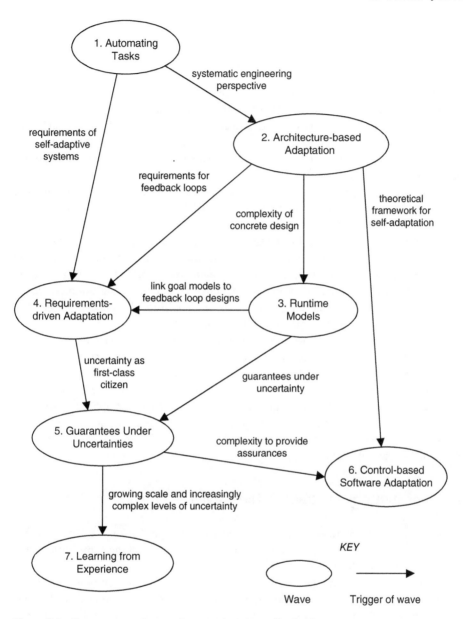

Figure 2.1 Seven waves of research on engineering self-adaptive systems

The fourth wave, *Requirements-driven Adaptation*, is triggered by the need to consider requirements of self-adaptive systems as first-class citizens (from waves one and two) and link the requirements to feedback loop designs (from wave three). The fourth wave puts the emphasis on the requirements that need to be solved by the managing system and how these requirements drive the design of the managing system. A distinction can be made between requirements that accommodate the adaptation concerns (and need to be translated to operational adaptation goals) and requirements about the functionality of the managing system, in particular its correct realization.

The fifth wave, *Guarantees Under Uncertainty*, is triggered by the need to deal with uncertainty as a first-class citizen when engineering self-adaptive systems (from wave four) and how to mitigate this uncertainty and guarantee the adaptation goals (from wave three). The fifth wave is concerned with providing trustworthiness for self-adaptive systems that need to operate under uncertainty. An important aspect of this wave is the collection of evidence that a self-adaptive system has realized its adaptation goals. This evidence can be provided offline (i.e. not directly controlled by the running system) and complemented online (i.e. under control of the running system).

The sixth wave, *Control-based Software Adaptation*, is triggered by the complexity of providing assurances (from wave five) and the need for a theoretical framework for self-adaptation (from wave two). The sixth wave is concerned with exploiting the mathematical basis of control theory for designing self-adaptive systems and analyzing and guaranteeing key properties. Central aspects of this wave are the definition of adaptation goals, the selection of a controller, and the identification of a model of the managed system that is used by the controller. A particular challenge is understanding the relationship between quality requirements, adaptation goals, and traditional controller properties, which is important for providing guarantees for self-adaptive systems.

Wave seven, *Learning from Experience*, is triggered by the growing scale of systems and increasingly complex levels of uncertainty (from wave five). The seventh wave is concerned with exploiting machine learning techniques to support different functions that need to be realized by a managing system. Examples are the use of learning techniques to keep a runtime model up-to-date, to reduce very large search spaces of adaptation options, learning to determine the impact of adaptation decisions on the goals of systems, and sorting adaptation options by predicting the values of their quality properties in order to support efficient decision-making for adaptation.

2.2   Contributions Enabled by the Waves

Table 2.1 summarizes the relevant aspects of the state-of-the-art before each wave with a motivation for the wave, the topic that is studied in each wave, and the contributions that are enabled by each of the waves.

The table summarizes how the subsequent waves have triggered each other, contributing complementary knowledge on the engineering of self-adaptive systems. At the time of writing, waves W1 to W4 are relatively stable and have contributed a substantial body of knowledge to the field. Wave W5 is in an active stage, but this wave has already produced substantial consolidated knowledge. Waves W6 and W7 are relatively new, and the knowledge consolidated in these waves is still limited.

2.3   Waves Over Time with Selected Work

Figure 2.2 gives a schematic overview of when each of the seven waves of research in the field of self-adaptive systems emerged over time. The time window per wave represents

Table 2.1 Summary of the state-of-the-art before each wave with motivation, topic of each wave, and the contributions enabled by each wave (or *expected to be enabled* for active waves).

Wave	State of the art before wave and motivation for the wave	Topic of the wave	Contributions (to be) enabled by the wave
W1	System management done by human operators is a complex and error prone process	Automation of management tasks	Systems manage themselves based on high-level objectives; understanding of the basic functions of self-adaptation
W2	Motivation for self-adaptation acknowledged; need for a principled engineering perspective to define self-adaptive systems and reason about adaptation at runtime	Architecture perspective on self-adaptation	Architecture as driver for the engineering of self-adaptive systems; central role of architectural models to reason about adaptation at runtime
W3	Architecture principles of self-adaptive systems understood; concrete realization is complex	Model-driven approach extended to runtime to realize self-adaptation	Different types of runtime models as key elements to define feedback loops and support runtime decision-making
W4	Design of feedback loops well understood; requirements problem they intend to solve is implicit	Requirements for self-adaptive systems	Languages and formalisms to specify requirements for self-adaptive systems and their operationalization
W5	Mature solutions for engineering self-adaptive systems, but uncertainty handled in ad-hoc manner	The role of uncertainty in self-adaptive systems and how to mitigate it	Use of formal techniques (at runtime) to guarantee adaptation goals under uncertainty
W6	Engineering of MAPE-based self-adaption well understood, but solutions are often complex	Applying principles from control theory to realize self-adaptation	Control theory as a basis for the design and formal analysis of self-adaptive systems
W7	Growing scale of systems and increasingly complex levels of uncertainty they face	The use of learning techniques to support self-adaptation	Learning techniques used to support the effectiveness of different adaptation functions

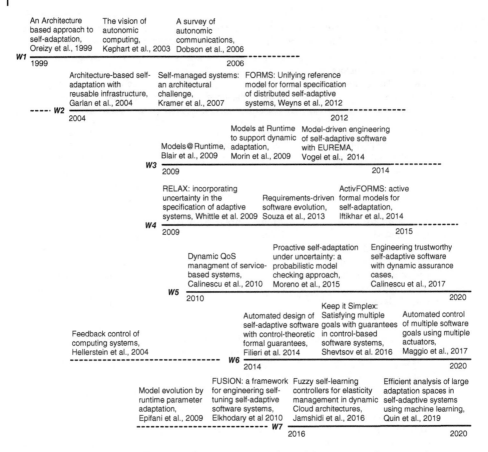

Figure 2.2 Main periods of activity of each wave over time with representative research papers.

the main period of research activity during a wave. The papers associated with each wave are either a key paper that triggered the research of the wave or papers that characterize the research for that wave. For Waves 6 and 7, which are still in an early stage, there were important precursor papers a few years before the research activities of these waves effectively took off.

2.4 Summary

Over the past two decades, the research in the field of self-adaptation went through seven waves. These waves put complementary aspects of engineering self-adaptive systems into focus. The waves highlight trends of interest in the community that together have produced the current body of knowledge in the field.

The root wave, *Automating Tasks*, deals with delegating complex management tasks of software-intensive systems from human operators to machines. *Architecture-based Adaptation* laid the basis for a systematic engineering approach for self-adaptive systems and the means to reason about self-adaptation during operation. The first two waves put the focus on the primary driver for self-adaptation and the fundamental principles of engineering self-adaptive systems.

The wave *Runtime Models* emphasizes the central role of models as first-class representations of key elements of a self-adaptive system to realize concrete feedback loops. *Requirements-driven Adaptation* put the requirements problem of self-adaptation on the agenda of the community. These two waves put the focus on core aspects of the concrete realization of self-adaptive systems.

Guarantees Under Uncertainty deals with the dichotomy between uncertainties on the one hand and guarantees for adaptation goals on the other hand. *Control-based Software Adaptation* aims at exploiting the mathematical framework of control theory to provide a foundation for self-adaptation. These two waves put the focus on uncertainties as a key driver for self-adaptation and how to mitigate them.

Finally, the most recent wave, *Learn from Experience*, aims to deal with the growing scale of systems and increasingly complex levels of uncertainty by using machine learning techniques.

The results obtained from subsequent waves have been the trigger for new waves. This way the waves have contributed complementary knowledge on the engineering of self-adaptive systems. The first four waves have reached a relatively stable status now, which has resulted in a substantial body of knowledge. The last three waves are the subject of intensive research in the community today; these waves are actively producing new insights and contributing knowledge in the field.

2.5 Bibliographic Notes

Pioneering work in the first wave, which was centered on automating tasks, was performed by P. Oreizy et al. who motivated the need for systems to self-adapt in order to achieve dependability [151]. J. Kephart and D. Chess introduced the notion of autonomic computing, referring to computing systems that can manage themselves given a set of high-level objectives, taking inspiration from biological systems [112]. S. Dobson et al. emphasized the central role of feedback loops in autonomic systems, focusing on communication and networks that automatically seek to improve their services to cope with change [65].

J. Kramer and J. Magee argued for an architectural perspective to self-adaptation, as this provides the required level of abstraction and generality to deal with the challenges self-adaptation poses [121]. D. Garlan et al. introduced the principles of architecture-based adaptation, which relies on the use of an architectural model to reason about adaptation [81], aligned with [151]. With Rainbow, a concrete realization of architecture-based adaptation, the authors made a seminal contribution to the second wave. The FORMS reference model defined a precise vocabulary for describing and reasoning about the key architectural characteristics of self-adaptive systems [206].

Pioneering work in the third wave was performed by G. Blair et al. who introduced the notion of models at runtime to manage the complexity of developing adaptation mechanisms [31]. Morin et al. turned the concept of a runtime model into a concrete realization that is centered on variability management to handle changing operating conditions, leveraging principles of dynamic software product lines [145]. T. Vogel and H. Giese went a step further by keeping the designed feedback loop models alive at runtime and directly executing these models to realize self-adaptation [195].

J. Whittle et al. performed foundational work in the fourth wave by incorporating uncertainty in the specification of requirements for self-adaptive systems [214]. V. E. Silva Souza et al. introduced awareness and evolution requirements in answer to the question of what requirements problem a feedback loop is intended to solve [184]. U. Iftikhar et al. provided guarantees for the correct behavior of the feedback loop with respect to a set of correctness properties by employing direct execution of formally verified models of the feedback loop [101].

Groundbreaking work in the fifth wave was performed by R. Calinescu at al. who exploited runtime quantitative verification to enable service-based systems to achieve their quality of service requirements by dynamically adapting to uncertainty in the system state, environment, and workload [39]. G. Moreno et al. applied probabilistic model checking at runtime to enable a system to make adaptation decisions with a look-ahead horizon, taking adaptation latency into account [144]. ENTRUST was the first methodology for the systematic engineering of trustworthy self-adaptive software, which combined design-time and runtime modeling and verification, with industry-adopted assurance processes based on assurance cases [42].

A seminal book of J. Hellerstein et al. outlined the potential for applying the principles of control theory to realize self-adaptive computing systems [94]. A. Filieri et al. performed foundational work in the sixth wave by developing a methodology for automatically constructing a dynamic model of a software system and a suitable controller for managing its adaptation requirements [71]. Leveraging this work, S. Shevtsov et al. developed SimCA, an automated control-based approach to software adaptation that can handle multiple goals [176]. M. Maggio et al. pushed the application of automated control-based adaptation further by applying model-predictive control [134].

Pioneering work in the seventh wave was performed by I. Epifani et al. who applied a Bayesian learning approach to keep runtime models up-to-date during execution [68]. A. Elkhodary et al. introduced the FUSION framework, which learns the impact of adaptation decisions on the system's goals to deal with unanticipated conditions and enhance the efficiency of run-time analysis [67]. P. Jamshidi et al. applied fuzzy Q-learning to enable an auto-scaler to automatically learn and update its scaling rules [107]. Finally, F. Quin et al. applied an online classifier to reduce large adaptation spaces, supporting efficient runtime analysis of large-scale self-adaptive systems [159].

An initial description of the first six waves outlined in this chapter was given in [198].

3

Internet-of-Things Application

An important area for applying self-adaptation that emerged around 2010 is the Internet-of-Things (IoT). In this chapter, we introduce an IoT application for building security monitoring, called DeltaIoT. We will use this application as a running example in the following chapters to illustrate the subsequent waves of research in the field of self-adaptive systems.

> *LEARNING OUTCOMES*
>
> - To understand the setup and basic technical characteristics of DeltaIoT.
> - To understand the main sources of uncertainties in DeltaIoT and explain their potential impact.
> - To explain the quality requirements of DeltaIoT and the impact of uncertainties on these.
> - To discuss manual management of the IoT network and motivate the need for self-adaptation.

3.1 Technical Description

DeltaIoT consists of a collection of 15 nodes that are connected through a wireless network. Fourteen of these are battery-powered IoT motes that are strategically placed over a geographical area as shown in Figure 3.1. The motes are equipped with sensors to provide access control to certain buildings (RFID sensor), to monitor motion in buildings and other areas (passive infrared sensor), and to sense the temperature. The sensor data of all the motes is relayed to the fifteenth node, which is an IoT gateway that is deployed at a central monitoring facility. Security staff can monitor the status of buildings from the monitoring facility and take appropriate action whenever unusual behavior is detected.

DeltaIoT uses multi-hop wireless communication. As shown in Figure 3.1, each IoT mote in the network relays its sensor data to the gateway. IoT motes that are farther away from the gateway have to relay their sensor data via intermediate IoT motes.[1] DeltaIoT uses time

---

1 We say that the sending mote is a child from the viewpoint of the receiving mote, and the receiving mote is a parent from the viewpoint of the sending mote.

An Introduction to Self-Adaptive Systems: A Contemporary Software Engineering Perspective,
First Edition. Danny Weyns.
© 2021 John Wiley & Sons Ltd. Published 2021 by John Wiley & Sons Ltd.

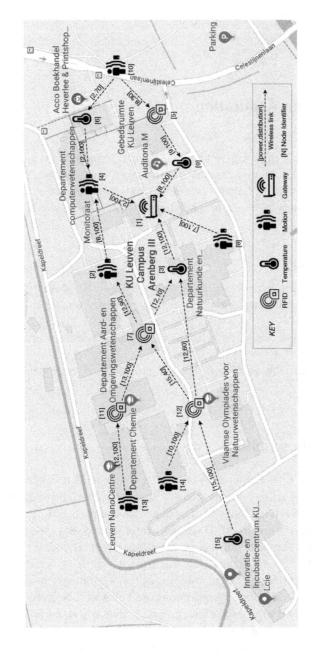

Figure 3.1 Geographical deployment of the DeltaIoT network

synchronized communication; concretely, the communication in the network is organized in cycles, each cycle comprising a fixed number of communication slots. Each slot defines a sender mote and a receiver mote that can communicate with one another. The communication slots are divided fairly among the motes. A communication cycle typically takes orders of minutes. For instance, the system can be configured with a cycle time of 9.5 minutes (570 seconds) with each cycle comprising 285 slots, each of 2 seconds. The slots in the first 8 minutes are allocated to the motes to communicate their sensor date to the gateway (i.e. downstream); the remaining 1.5 minutes are allocated to the gateway to communicate messages to the motes – for instance messages with information for the motes on how to adapt their network settings (upstream).

Each mote is equipped with three queues: *buffer* collects the packets produced by the mote, *receive-queue* collects the packets from the mote's children, and *send-queue* queues the packets to be sent to the parent(s) during the next cycle. The size of the *send-queue* is equal to the number of slots that are allocated to the mote for communication during one cycle. Before communicating, the packets of the *buffer* are first moved to the *send-queue;* the remaining space is then filled with packets from the *receive-queue*. Packets that arrive when the *receive-queue* is full are discarded (i.e. *queue loss*).

DeltaIoT offers a management interface, which is accessible via the gateway. System operators can use this interface to collect data about the network and set the network parameters, as schematically illustrated in the architectural model of Figure 3.2.

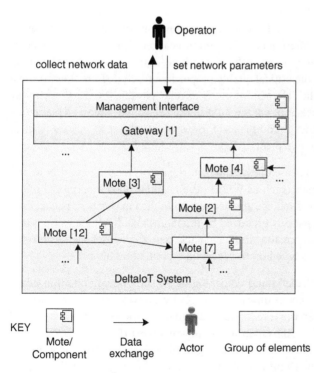

Figure 3.2 Part of the DeltaIoT network architecture with Gateway and Management interface

The management interface provides information about the traffic generated by each mote in each cycle (number of packets), the energy consumed (in Coulomb, or C, typically on the order of 10's of mC per cycle), the Signal-to-Noise ratio (SNR) for each link (in decibel, or dB, typically in the range of 10 to -40), the settings of the transmission power that a mote used to communicate with each of its parents (in a range from 0 to 15), and the distribution factor per link being the percentage of the packets sent by a mote over the links to each of its parents (0 to 100%). The interface also provides access to statistical information about the Quality of Service (QoS) of the overall network for a given period. QoS includes data about packet loss (the fraction of packets lost [0…1]), energy consumption (Coulomb), and latency of the network (fraction of the cycle time that packets remain within the network [0…1] – 0 means all packets are delivered in the cycle that they are generated; 1 means all packets are delivered in the cycle after the one in which they are generated).

In addition, the management interface enables operators to set the parameters for each mote and each link of the network. This includes the transmission power to be used by a mote to communicate via the link (from 0 being the minimum level of energy to 15 being the maximum level) and the distribution factor for the link (0 to 100%). Finally, operators can reset the network settings (transmission power and distribution factors) to predefined default values at any time.

3.2 Uncertainties

IoT applications are expected to remain fully operational for a long time on a set of batteries (typically multiple years), while offering reliable communication with low latency. To guarantee these quality properties, the motes of the network need to be optimally configured. Two key factors that determine the critical quality properties of DeltaIoT are the transmission power of the motes and the selection of paths to relay packets toward the gateway (i.e. the distribution of the packets sent via the links to the respective parents). Guaranteeing the required quality properties is complex as the system is subject to various types of uncertainties. Here, we consider two primary types of uncertainty:

1. *Network interference and noise:* Due to external factors, such as weather conditions and the presence of wireless signals, such as WiFi in the neighborhood, the quality of the communication between motes may be affected, which in turn may lead to packet loss.
2. *Fluctuating traffic load:* The packets produced by the motes may fluctuate in ways that are difficult to predict; for instance, the number of packets produced by a passive infrared sensor at some point in time is based on the detected motion of humans.

As an example, the graph on the left hand side of Figure 3.3 shows the fluctuating values of the SNR (in *dB*) of one of the communication links between two motes over time. The *SNR* represents the ratio between the level of desired signal and the level of the undesired signal, i.e. noise, which comes from the environment in which the IoT system operates. The higher the interference, the lower the SNR, resulting in higher packet loss. The graph on the right shows the frequency distribution of the same data as the graph on the left, which has a normal distribution.

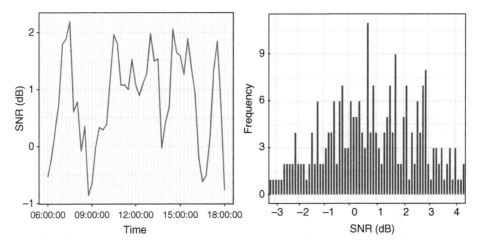

Figure 3.3 Uncertainty due to interference for one of the communication links in Figure 3.1.

3.3 Quality Requirements and Adaptation Problem

The quality requirements for DeltaIoT defined by the stakeholders are:

R1: The average packet loss per period of 12 hours should not exceed 10%.
R2: The energy consumed by the motes should be minimized.
R3: The average latency of packets per 12 hours should be less than 5% of the cycle time.

The key problem of DeltaIoT is how to ensure the quality requirements of the system regardless of the uncertainties in the IoT network caused by network interference and the fluctuating traffic load of packets. A typical approach used in practice to deal with such uncertainties in IoT applications is to over-provision the network. In this approach, the transmission power of the links is set to maximum, and all packets transmitted by a mote are copied to all its parents. While such a conservative approach may result in low packet loss, the cost is high energy consumption, resulting in a reduced lifetime of the network. Once the system is in operation, operators use the management interface to track the system behavior and fine-tune the initial settings based on observations of the network, aiming to reduce the overall energy consumption. However, manual intervention is a costly, slow, and error-prone activity. By enhancing DeltaIoT with self-adaptation capabilities, the system will automatically track the uncertainties at runtime and use up-to-date information to find and adapt the settings of the motes such that the system complies with the quality requirements.

3.4 Summary

IoT is an important emerging field for applying self-adaptation. DeltaIoT is a representative example IoT application in the domain of building security monitoring.

DeltaIoT consists of 14 geographically distributed motes equipped with sensors that communicate packets of sensor data to a gateway via a wireless network.

DeltaIoT is subject to network interference and noise from the environment. Another uncertainty is fluctuating traffic load in the network as motes only communicate data when necessary. These uncertainties are characteristic for wireless IoT applications.

The quality requirements of DeltaIoT are keeping the average packet loss and latency below a given threshold while minimizing the energy consumed by the motes. These are representative quality requirements for IoT applications.

A key problem of DeltaIoT and IoT applications in general is how to ensure the quality requirements despite the uncertainties the network is subjected to.

A typical approach applied in practice is to over-provision the network – i.e. to set the packet communication power at the motes to maximum and send to duplicates of packets to all parents. While this approach keeps the packet loss low, the cost is high energy consumption. Therefore, operators manually tweak the network settings based on observations and trial and error.

Manual management of the IoT network is costly, slow, and error-prone. Self-adaptation offers an opportunity to tackle these problems by automating the management of the network.

3.5 Exercises

3.1 Simulate DeltaIoT network: level D
Setting. Consider the standard setup of the DeltaIoT network with two requirements:
R1: The average packet loss over 100 cycles should not exceed 10 %.
R2: The energy consumed by the motes should be minimized.
We use the DeltaIoT simulator and manually apply different settings of the network through the management interface. We are interested in the effects of different settings on the energy consumed by the motes and the reliability of the messages transmitted to the gateway.
Task. Download the DeltaIoT artifact and install the simulator. Define 10 different settings of the network by varying the power settings and the distribution factors of motes. Run the simulator for each setting for a period of 100 cycles. Collect the data of energy consumption and packet loss for each setting. Analyze the data and assess the results in light of requirements R1 and R2. Interpret your results.
Additional material. See the DeltaIoT artifact website [103] The latest version of the simulator can be downloaded from the DeltaIoT website [213].

3.2 Literature review of self-adaptation in IoT: level W
Setting. The Internet-of-Things is an emerging area for the application of self-adaptation. We want to understand the state of the art of self-adaptation applied to IoT.
Task. Perform a systematic literature review on the use of self-adaptation techniques in the area of the Internet-of-Things. Identify the motivations for why self-adaptation

has been used in IoT, the adaptation techniques that have been used with their purpose, how the techniques have been used, and how they have been evaluated. Analyze the collected data and draw conclusions. Identify open challenges in the application of self-adaptation to IoT.

Additional material. For guidelines on performing systematic literature reviews, see [118].

3.6 Bibliographic Notes

The Internet-of-Things is an emerging domain for the application of self-adaptation. We illustrate this with three research results focusing on different challenges. G. Sankar Ramachandran et al. presented Dawn, an optimization approach for component-based IoT systems that automatically extracts and enforces bandwidth requirements from component compositions [162]. T. Bures et al. modeled IoT systems as ensembles and studied their ability to self-adapt [37]. J. Beal et al. studied stability in large-scale IoT systems that use a computational model that relies on the field calculus [19].

U. Iftikhar et al. introduced the DeltaIoT artifact [102]. The artifact has been used to evaluate several new research results on self-adaptation, for instance to measure the degrees of satisfaction of tradoffs between non-functional requirements in self-adaptive systems [66], and to study the application of different machine learning techniques to reduce large adaptation spaces in self-adaptive systems [159]. For an industrial evaluation of DeltaIoT, see [211].

4

Wave I: Automating Tasks

When Paul Horn introduced IBM's manifesto on autonomic computing to the National Academy of Engineers at Harvard University in 2001, he pointed to the rising complexity of computing systems in the growing size of code, the increasing heterogeneity of software environments, and the progressing interaction and integration of system components across company boundaries into the Internet. These new levels of complexity introduced severe manageability problems in installing, configuring, operating, and maintaining computing systems that went beyond the capabilities of even the most skilled system administrators.

Around the same time, researchers in the USA and Europe raised the issue that existing software engineering approaches were not delivering their full premise. Although solutions for systematic and principled design and deployment of applications existed, such as object-oriented analysis and design and high-level programming languages, these solutions were not retaining the full plasticity that would allow software to be easily modified in the field. This modifiability is essential to accommodate resource variability and changing user needs, and handle system faults, without requiring system restart or downtime.

The manageability problems of complex software systems were the primary motivation for self-adaptation and triggered the first wave: Automating Tasks. The ability of a software system to be easily modifiable throughout its lifetime is a basic system property to enable automation of tasks.

In this chapter, we zoom in on autonomic computing and the automation of tasks. We define the core functions of self-adaptation, we put adaptation management in context and we connect it with evolution management.

LEARNING OUTCOMES

- To explain and illustrate the principle of autonomic computing to automate mainte-nance tasks.
- To explain and illustrate the essential types of maintenance tasks for automation.
- To explain the primary functions that are required to automate tasks and realize self-adaptation.
- To analyze a concrete self-adaptive application and identify its primary functions.
- To explain adaptation management and evolution management and their relation-ship.

An Introduction to Self-Adaptive Systems: A Contemporary Software Engineering Perspective,
First Edition. Danny Weyns.

4.1 Autonomic Computing

Unlike today, in the late 1990s through the late 2000s, installing, configuring, and maintaining software systems required manual intervention. For instance, the most common way to get new software was to purchase a floppy disk or CD-ROM, insert the disc, and walk through the installation instructions. For large complex computing systems, applying such manual procedures led to growing manageability problems for system operators.

To tackle these problems, IBM introduced a new vision on engineering such systems that they coined *autonomic computing*. The principal idea of autonomic computing was to free administrators from system installation, operation, and maintenance by letting computing systems manage themselves based on high-level objectives provided by administrators.

The idea of autonomic computing is inspired by our autonomic nervous system, which seamlessly governs our body temperature, heart beat, breathing, etc. Similarly, autonomic systems should seamlessly adapt their operation to dynamic workloads and changing external conditions. To that end, the autonomic system continually monitors itself, analyzes the situation, and reconfigures itself when necessary to maintain its high-level objectives.

The capabilities of an autonomic computing system rely on a *feedback loop*[1] that collects data from the system and its environment and acts accordingly to realize its high-level objectives. This automation reduces the manual effort and time required to respond to critical situations, resulting in systems that operate more effectively in the face of changing operating conditions.

Example We illustrate the use of a feedback loop to automate a task in DeltaIoT. The left side of Figure 4.1 shows the architecture of the DeltaIoT system enhanced with a feedback loop. The objectives for the feedback loop are set by an operator – for instance, keep the average packet loss below a given threshold while minimizing the energy consumption of the motes. The task of the feedback loop is to adapt the network settings automatically such that these objectives are achieved regardless of changing operating conditions such as changing levels of interference along the links between motes of the network.

To realize this task, the feedback loop exploits the management interface to track various parameters of the network, including the current settings of the motes (transmission power per link, distribution of messages to parents), the current values of uncertainty parameters (number of messages generated by the motes and the SNR for each link), and the current qualities of the network (packet loss of the network, the energy consumed, and the latency of the network).

Figure 4.1 top right shows the settings of Mote [12] at a particular point in time. The transmission power for the link to Mote [3] is set relatively high, while the power for the link to Mote [7] is set rather low (represented by the sliders in the figure). In the given

---

1 IBM coined the term "autonomic element" to emphasize the self-governing nature of autonomic computing systems, seeking inspiration from social, economic, and biological systems. Autonomic systems could be hierarchical in nature and act as distributed or multi-agent systems. Aligned with the conceptual model of a self-adaptive system, we consistently use the term "feedback loop" in this book to refer to the adaptation logic of a self-adaptive system that deals with one or more adaptation goals. We use the term feedback loop as a unified concept for different terms that have been used in the literature, such as autonomic element, adaptation engine, and plant controller.

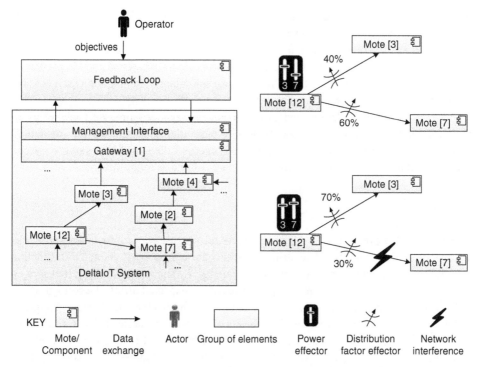

Figure 4.1 Left: DeltaIoT system extended with feedback loop. Top right: excerpt of the network with settings of Mote [12]; bottom right: adapted settings with interference on the link to Mote [7].

configuration, 60 % of the packets of Mote [12] (received from its children and generated by the mote itself) are sent to Mote [7]; the remaining 40 % are sent to Mote [3] (this is represented by the control valves in the figure).

Figure 4.1 bottom right shows a scenario where unexpected interference occurred along the link from Mote [12] to Mote [7]. If the system would use the initial settings of Mote [12], this may result in a significant loss of packets transmitted via this link.

When the feedback loop detects the interference, it will adapt the network settings to ensure the high-level objectives. Concretely, the feedback loop took two actions in this scenario: first, it increased the power setting of the link to Mote [7], and second, it reduced the packets sent to Mote [7] to 30 %; the remaining 70 % are sent to Mote [3].

By increasing the transmission power of Mote [12] for packets sent via the link to Mote [7], the SNR along this link will be improved reducing packet loss. By adjusting the distribution of packets, more packets will be routed to the gateway via Mote [3]. Note that not all packets are sent to Mote [3] as this may overload the queues of Mote [3].

4.2 Utility Functions

When an autonomic system needs to make an adaptation decision, there are usually multiple options to select from. One fundamental mechanism to specify the preference of one option over another is a utility function. Utility functions are well-known in other fields such as economics and social sciences.

In the context of self-adaptation, utility functions usually compute the expected utility of choices among different options, i.e. a preference ordering over the choice set. In particular, a utility function provides an objective function for self-adaptation, mapping each possible configuration of the system to a real scalar value.

The expected utility is determined by a function that encodes a measure of the "goodness" (or "badness") of the result of choosing a particular configuration for adaptation. Hence, such a function captures the utility preferences of the stakeholders of the system. The expected utility U_c for a configuration c can be defined as follows

$$U_c = \sum_{i=1}^{n} w_i \cdot p_i \qquad (4.1)$$

with p_i the utility preference of the stakeholders for property i of n properties and w_i the relative weight for property i. The utility preference for a property is defined by a function that maps property values to a value in the domain [0...1]. The relative weights that sum up to 1 express the relative importance of properties in the expected utility of a configuration.

Consider as an example the service-based health system that we introduced in Chapter 1 with two quality properties that are the subjects of adaptation: service failures and service cost. Assume that the stakeholders assign a utility preference of 1 to failure rates below 1 %, 0.3 to failure rates between 1 % and 2 %, and 0 to failure rates above 2 %. Assume that the stakeholders assign a utility preference of 1 to costs below $ 5, 0.7 to costs between $ 5 and $ 10, 0.5 to costs between $ 10 and $ 15, and 0 to costs above $ 15. These utility preferences are graphically illustrated in Figure 4.2.

Furthermore, assume that the stakeholders give a weight of 0.7 to failure rate and a weight of 0.3 to cost. Consider now a situation where a new service configuration needs to be selected from two possible options: the first configuration C1 has a failure rate of 1.2 % and a cost of $ 11 while the second configuration C2 has a failure rate of 0.9 % and a cost of $ 16. The expected utilities for the failure rate and cost of the two configurations are marked with C1 and C2 in Figure 4.2. The expected utility of service configurations can

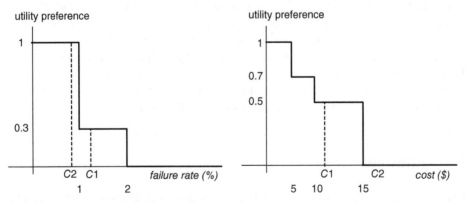

Figure 4.2 Utility preferences for failure rate and cost in the example.

then be computed as

$$U_c = w_{\text{failure\_rate}} \cdot p_{\text{failure\_rate}} + w_{\text{cost}} \cdot p_{\text{cost}} \qquad (4.2)$$

$$U_{C1} = 0.7 \cdot 0.3 + 0.3 \cdot 0.5 = 0.36 \qquad (4.3)$$

$$U_{C2} = 0.7 \cdot 1.0 + 0.3 \cdot 0.0 = 0.70 \qquad (4.4)$$

Hence, in this situation, service configuration C2 with the highest expected utility will be preferred over configuration C1 and will be selected for adaptation.

4.3 Essential Maintenance Tasks for Automation

The primary objective of autonomic computing is to free system administrators from the complexity of operating and maintaining systems that are subject to change. Autonomic computing systems are expected to provide their service to clients as required, 24/7. Often, a distinction is made between four essential types of maintenance tasks than can be automated using autonomic computing. These tasks differ in the type of problems they pose and the specifics of the automation solutions they require.

4.3.1 Self-Optimization

Complex systems are subject to resource constraints, such as limited memory, computational power, bandwidth, and energy. The cost of using third-party resources adds an additional constraint. Ignoring these constraints may jeopardize the sustainability of the system and/or the profitability of the business. Hence, optimal usage of the constrained resources is a necessity.

> Self-optimization is the capability of a system to continuously seek opportunities to optimize its resources' usage, while providing the required quality objectives.

We illustrate self-optimization for a scenario in DeltaIoT as shown in Figure 4.3. Assume that in the given setting, the feedback loop deployed at the gateway (see Figure 4.1) observes that the packet loss goal (<10%) and the latency goal (<5% of the cycle time) are met. Nevertheless, the feedback loop may check whether there is any opportunity to adapt the current configuration to optimize the energy that is consumed by the motes.

To that end, the feedback loop identifies a set of possible configurations of the network (based on the ranges of possible settings of the transmission power per link, and the distributions of packets sent by motes with multiple parents, i.e. the distribution factors per link). For each of the configurations, the feedback loop predicts what the expected packet loss, latency, and energy consumption will be. Such predictions can be determined using different techniques. One such technique is predictive modeling, which uses statistics to predict the expected value of a variable based on a model.

The table in Figure 4.3 summarizes the results for a sample of the possible configurations. The current setting of the system corresponds with *configuration 3*, which is marked with an arrow in the table. Although the prediction results show that both the packet loss and

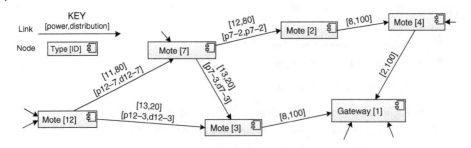

configuration	[power, distribution]				packet loss (%)	latency (%)	energy (C)
	[p12–7,d12–7]	[p12–3,d12–3]	[p7–2,d7–2]	[p7–3,d7–3]			
1	[11,60]	[13,40]	[12,100]	[13,0]	9	2	22
2	[11,70]	[13,30]	[12,90]	[13,10]	11	1	16
→ 3	[11,80]	[13,20]	[12,80]	[13,20]	9	4	24
4	[12,60]	[13,40]	[12,70]	[13,50]	8	11	18
5	[12,70]	[13,30]	[12,60]	[13,40]	9	3	19
6	[12,80]	[13,20]	[12,50]	[13,50]	13	2	16

Figure 4.3 Self-optimization scenario for DeltaIoT. The table shows the expected qualities for a sample of possible configurations. The arrow in this table points to the current configuration.

latency goals will be achieved with the current configuration, there are two other configurations (numbers 1 and 5) that are predicted to realize the goals, but with less required energy. Of these two, configuration number 5 is expected to consume the lowest amount of energy. Hence the feedback loop can exploit this opportunity and adapt the network settings accordingly.

This self-optimization scenario shows how the feedback loop continuously seeks opportunities to optimize the energy consumed by the motes in the network, while maintaining the required quality properties.

4.3.2 Self-Healing

Complex systems are prone to errors, such as bugs or faults in software or hardware. Such errors may lead to failures that can jeopardize normal system operation resulting in serious problems for users or customers. Manually identifying the root cause of a failure, diagnosing the situation, and resolving the problem can be labor intensive and expensive. Consequently, it is important that failures are detected quickly and that the problems are localized and resolved efficiently and effectively.

> Self-healing is the capability of a system to detect and recover from failures, in order to provide its required quality objectives, or degrade gracefully otherwise.

Figure 4.4 shows a scenario in DeltaIoT that illustrates self-healing. The feedback loop deployed at the gateway (see Figure 4.1) regularly checks whether any of the motes of the network has failed. To that end, different strategies can be applied. A classic mechanism is a heartbeat protocol where the motes periodically send a signal to the gateway to indicate their normal operation. An alternmative is a ping-echo protocol where the gateway sends

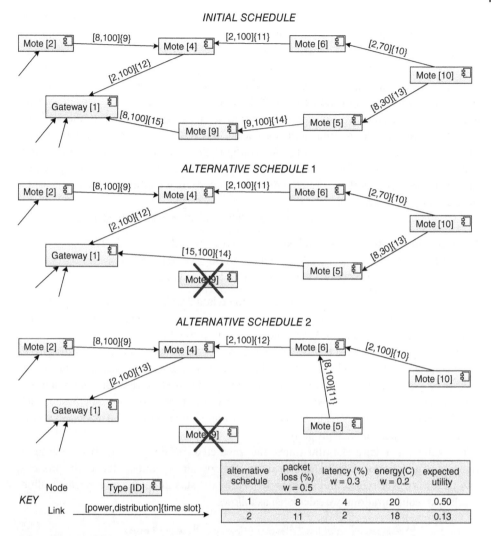

Figure 4.4 Self-healing scenario for DeltaIoT. The tables summarizes the results of the calculation of the expected utility for two alternative time-synchronization schedules after Mote [9] fails.

requests to the motes and waits a given time period for echo replies. To avoid unnecessary energy consumption, only the motes that are not actively producing data need to be checked.

When a mote fails, its sensor data will be lost. However, the failure may also affect the data transmission of other motes. Consider the initial configuration shown at the top of Figure 4.4. Since the communication in DeltaIoT is time scheduled, the communication in the network is organized in cycles, and connected motes are assigned a time slot in each cycle. The ordering of this schedule is indicated on the links with a number between curly brackets. If Mote [9] in the original configuration fails, not only will the packets generated by this mote be lost, but also the packets Mote [9] receives from Mote [5] and indirectly from Mote [10] via Mote [5] will no longer reach the gateway.

When the feedback loop detects the failure of Mote [9], it uses a model of the network to identify alternative time schedules to synchronize the communication among the motes such that the data of all active motes can reach the gateway. Two such schedules are shown in Figure 4.4. To select the best time-synchronization schedule from the alternatives, the feedback loop needs to determine the best option, taking into account the physical constraints of the network. Different techniques can be used to make such decisions. For instance, in the self-optimization scenario discussed above, the feedback loop uses rules that define thresholds for packet loss and latency. These rules determine the valid configurations. Among the valid configurations, the one with minimum energy consumption is selected for adaptation. Let us now consider an alternative strategy where the feedback loop uses the *expected utility* of the alternative configurations as a basis for making adaptation decisions.

For DeltaIoT, the utility preferences of the stakeholders may be defined as follows:

1. packet loss: utility = 1 if packet loss $\leq$5 %, 0.5 if >5 % and $\leq$10%, 0 if >10 %.
2. latency: utility = 1 if latency $\leq$2 %, 0.5 if >2 % and $\leq$5 %, 0 if >5 %.
3. energy: utility = 1 if energy $\leq$10 C, 0.5 if >10 C and $\leq$20 C, 0 if >20 C.

Let us assume that the stakeholders assign the relative weights to these three properties as follows: packet loss, being the most important property, is assigned a relative weight of 0.5, followed by latency 0.3, and then energy consumption 0.2.

To determine the expected utility of an alternative configuration, the feedback loop needs to be equipped with a mechanism that allows it to determine the expected values for the quality properties of interest. Various techniques can be used for this. One approach is to use parameterized models of each property, assign values to the parameters of these models based on the settings of each alternative configuration, and simulate the models to predict approximate values for each property.

The table in Figure 4.4 illustrates the predicted results for the two alternative time-synchronization schedules. Based on the predicted values for each property, and the utility preferences and weights defined by the stakeholders, the expected utilities for the two configurations are calculated as follows

$$U_c = w_{packet\_loss} \cdot P_{packet\_loss} + w_{latency} \cdot P_{latency} + w_{energy} \cdot P_{energy} \qquad (4.5)$$

$$U_1 = 0.5 \cdot 0.5 + 0.3 \cdot 0.5 + 0.2 \cdot 0.5 = 0.50 \qquad (4.6)$$

$$U_2 = 0.5 \cdot 0.0 + 0.3 \cdot 1.0 + 0.2 \cdot 0.5 = 0.13 \qquad (4.7)$$

Based on these results, the feedback loop will select time-synchronization schedule 1 and adapt the underlying network accordingly.

This self-healing scenario shows how the feedback loop continuously tracks failing motes in the network to recover by adapting the system, in order to maintain its required quality properties.

4.3.3 Self-Protection

Firewalls and intrusion-detection software offer some level of protection to complex systems. However, systems may remain exposed to a large attack surface, despite the use of

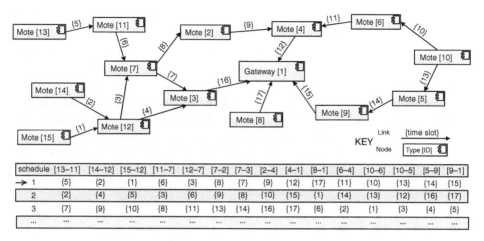

schedule	[13–11]	[14–12]	[15–12]	[11–7]	[12–7]	[7–2]	[7–3]	[2–4]	[4–1]	[8–1]	[6–4]	[10–6]	[10–5]	[5–9]	[9–1]
→ 1	{5}	{2}	{1}	{6}	{3}	{8}	{7}	{9}	{12}	{17}	{11}	{10}	{13}	{14}	{15}
2	{2}	{4}	{5}	{3}	{6}	{9}	{8}	{10}	{15}	{1}	{14}	{13}	{12}	{16}	{17}
3	{7}	{9}	{10}	{8}	{11}	{13}	{14}	{16}	{17}	{6}	{2}	{1}	{3}	{4}	{5}
...	...	...	...	...	...	...	...	...	...	...	...	...	...	...	...

Figure 4.5 Self-protection scenario for DeltaIoT. The tables shows a subset of time-synchronization schedules. The feedback loop can switch between these schedules to anticipate jamming attacks. The arrow refers to the schedule currently in use.

such tools. For complex systems, it becomes particularly hard for humans to detect malicious attacks, decide what measures to take to protect systems, and avoid cascading effects. Hence, it becomes crucial for systems to protect themselves against attacks and apply strategies to anticipate problems.

> Self-protection is the capability of a system to defend against malicious attacks and anticipate problems they may cause, in order to deliver its required quality objectives.

We illustrate self-protection for a scenario of DeltaIoT that is subject to jamming attacks. In a jamming attack, an attacker scans the transmissions of messages along the links in a network to detect patterns. The attacker then exploits this information to start transmitting spurious messages simultaneously with the regular network transmissions to corrupt the network traffic. Malicious jamming attacks that are applied in either a selective or random manner are very difficult to detect.

Figure 4.5 shows the network topology of DeltaIoT with a particular time-synchronization schedule as indicated by the numbers in the curly brackets marked on the links.

The feedback loop that is deployed at the gateway has access to traffic profiles, which serve as a baseline for normal traffic behavior. Such profiles can be generated based on stakeholder input, based on data collected in a trusted setting, or based on historical data. Furthermore, the feedback loop has a set of alternative time-synchronization schedules that can be used to organize the communication in the network. The table in Figure 4.5 shows a small subset of such schedules.

During operation, the feedback loop tracks the transmission of packets in the network, compares the actual traffic with the baseline, and raises an alert based on traffic variations that may indicate possible jamming attacks. When an alert is generated, the feedback loop selects an alternative time-synchronization schedule and adapts the underlying network accordingly. The alternative schedule may be selected randomly from the set of alternative schedules, or it may be determined based on some metric, for instance select a schedule that maximizes the time difference between the time slots of the links in the network. The new setting will remove the jamming of traffic.

This example realizes a reactive adaptation strategy where the system only adapts when suspicious behavior is detected. An alternative, preventive strategy would be to switch the time-synchronization schedule randomly from time to time. This will make it difficult for attackers to jam the network. However, a preventive strategy creates overhead for the underlying network, which needs to switch the time schedule regularly, and such schedules may also be sub-optimal with respect to the required qualities of the system.

The self-protection scenario shows how the feedback loop continuously scans the network traffic and adapts the communication schedule of the network when suspicious behavior is detected, in order to maintain the system's quality properties.

4.3.4 Self-Configuration

Large-scale systems regularly need to integrate new elements. Manually installing and configuring new elements of such systems during operation is complex, time consuming, and error prone. Hence, automated and seamless integration of new elements according to high-level objectives, without interrupting normal operation of the system, is important.

> Self-configuration is the capability of a system to automatically integrate new elements, without interrupting the system's normal operation.

We illustrate self-configuration for a scenario in DeltaIoT as shown in Figure 4.6. The figure on the left shows a part of the initial network configuration with a time-synchronization schedule.

To support self-configuration, the feedback loop deployed at the gateway (see Figure 4.1), together with the motes of the network, needs to implement a mote discovery mechanism. When a new mote is activated, it regularly broadcasts some basic information, such as its identifier, location, and the expected traffic load the mote will generate. The feedback loop

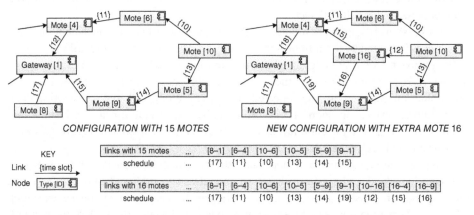

Figure 4.6 Self-configuration scenario for DeltaIoT. Left: initial configuration with a time-synchronization schedule. Right: configuration with schedule after Mote [16] joined the network.

on the other hand is equipped with a monitor that listens to the broadcast messages of newly arriving motes.

When the feedback loop discovers a new mote, it uses the data it received together with a model of the current network configuration to determine how to integrate the mote into the network. Applied to the scenario in Figure 4.6, when the feedback loop discovers Mote [16], it determines the motes that are in communication range of the new mote and uses this information to assign parents and children to the new mote. Among the different possible configurations, the feedback loop will select the most suitable configuration based on the quality objectives of the network. Different techniques can be used to make this decision – for instance, based on simulations of models of the system that allow the expected values for the different quality properties to be determined. In the scenario, the feedback loop assigns the role of parent of the new mote to Mote [10] and the role of child to Mote [4] and Mote [9]. In an alternative configuration, Mote [16] has no children and directly communicates the data it generates to the gateway. However, if Mote [16] is located far from the gateway, this configuration may require substantially more energy to transmit the data.

Once the new network topology is set, the feedback loop needs to adapt the time-synchronization schedule to match the new structure. In the example, three new time slots are instantiated for the connections of Mote [16] with its parent and child motes, and the time slot for the communication between Mote [9] and Mote [1] is changed (from Slot 15 to Slot 19). Finally, the feedback loop will adapt the current schedule of the underlying network to this new schedule. This completes the integration of Mote [16] into the network.

This self-organizing scenario shows how the feedback loop continuously monitors the activation of new motes and adapts the underlying time-synchronization schedule of the network accordingly in order to provide the required quality objectives of the IoT system.

4.4 Primary Functions of Self-Adaptation

The capabilities of an autonomic computing system rely on a feedback loop, which maps to the conceptual feedback loop of the managing system in the conceptual model of a self-adaptive system. We direct our attention now to the internals of the managing system. In particular, we identify the primary functions of a managing system to define a reference model. The reference model enables us to explain in broad terms how the elements of a self-adaptive system work together to accomplish their collective purpose.

A reference model is a commonly agreed decomposition of a known problem into *functional elements* that cooperatively solve the problem. The problem in this context is "how to realize self-adaptation?" To that end, a reference model defines divisions of functionality together with flows between the elements. Reference models consolidate experience; they characterize the maturity of a domain. In our case, this domain is the general domain of self-adaptive systems. It is important to note that a reference model is not the same as a reference architecture. A reference architecture is a reference model mapped onto *software elements* that cooperatively implement the functionality defined in the reference model. This mapping may or may not be one to one – i.e. a software element may implement a single function of the reference model, a part of a function, or several functions.

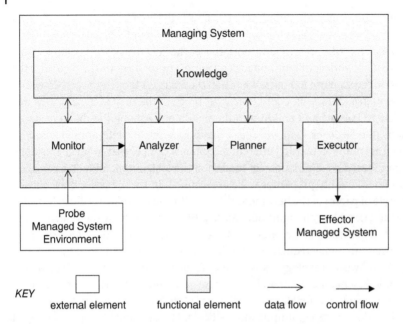

KEY

Figure 4.7 Reference model of a managing system.

Figure 4.7 shows the reference model of a managing system that realizes self-adaptation. The functional elements – Monitor, Analyzer, Planner, and Executor – realize the four basic functions of any self-adaptive system. These elements share common Knowledge. Hence the reference model of a managing system is often referred to as the *MAPE-K* model.[2]

In short, the functional elements of a managing system realize self-adaptation as follows. *Monitor* acquires data from the managed system and the environment through a probe, and processes this data to update *Knowledge* accordingly. *Analyzer* uses the up-to-date knowledge to determine whether there is a need for adaptation and if so it analyzes the options for adaptation. If adaptation is required, *Planner* selects the best adaptation option and puts together a plan that consists of one or more adaptation actions to adapt the system from its current configuration to the new configuration. The plan is then executed by *Executor* through an effector that adapts the managed system as planned.

MAPE-K provides a simple yet powerful reference model of a managing system. MAPE-K's power is the intuitive structure of the distinct functions that realize the feedback loop of a self-adaptive system. We now zoom in on the different functional elements of MAPE-K.

4.4.1 Knowledge

The knowledge function provides the knowledge and mechanisms that are required for the MAPE elements to work together and collaboratively realize self-adaptation. Knowledge

---

2 In their original work, IBM used the term "autonomic element" for a managing system and emphasized that such elements could be hierarchically structured or viewed as a multi-agent system. The reference model focuses on core functionalities, abstracting away from concrete mappings and compositions of functions onto architectural elements.

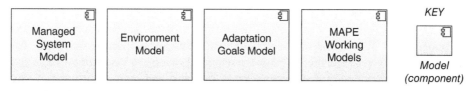

KEY

*Model
(component)*

Figure 4.8   Essential models managed by the knowledge of a managing system.

relies on four types of data, respectively related to the managed system, the environment, the adaptation goals, and the MAPE elements themselves. The knowledge function can be equipped with various mechanisms to support clients (MAPE elements) to read, write, and update the knowledge. A notification mechanism can be provided that notifies interested clients about certain changes that were made to the knowledge. In the event that multiple clients attempt to access knowledge simultaneously, mechanisms needs to be in place that avoid race conditions ensuring that the knowledge remains consistent.

In a concrete realization, the four types of knowledge are mapped to basic models, as illustrated in Figure 4.8.[3]

A managed system model represents the system that is managed by the managing system. Typically, an abstraction of the managed system is used that represents the parts of the managed system that are relevant to the realization of self-adaptation. For example, in DeltaIoT, the managed system model may represent the topology of the network with the settings of the motes and data about quality properties that are relevant to adaptation. By maintaining the managed system model, we say that a self-adaptive system becomes *self-aware*.

The environment model represents elements or properties of the environment in which the managed system operates. This model maintains representations of the parts of the world that are relevant to the realization of self-adaptation. For example, in DeltaIoT, the environment model may represent knowledge about the uncertainties, such as the traffic generated by motes and levels of interference of communication links. By maintaining an up-to-date model of the environment, we say that a self-adaptive system becomes *context-aware*.

The model of the adaptation goals maintains a representation of the objectives of the managing system. Research has shown that adaptation goals relate to quality properties of the managing system. The adaptation goals can be represented in different forms. A simple form is a set of rules, each rule representing a constraint on one quality property, together with an ordering in which the rules apply. Another form is a set of fuzzy goals, whose satisfaction is represented through constraints on the values. Fuzzy goals include 0 (false) and 1 (true) as extreme cases of truth, but also intermediate states of truth quantified as values between 0 and 1. Another common way to express preferences in terms of quality properties is by means of utility functions. As an example, in DeltaIoT, the adaptation goals may be represented as a set of rules that put constraints on the average packet loss and latency that are allowed over a period of time. By maintaining a model of the adaptation goals, we say that a self-adaptive system becomes *goal-aware*.

A MAPE working model represents knowledge that is shared between two or more MAPE elements. These models are typically domain-specific. Examples are working models that

---

3 Unless explicitly stated differently, we represent runtime models as software components.

are used by the MAPE elements to predict specific quality properties of the system. These models are usually parameterized, allowing the prediction of the qualities of different specified configurations by assigning different values to the parameters. Another working model is a plan that needs to be executed to adapt the managed system. In DeltaIoT, the managed system may maintain quality models for packet loss, latency, and energy consumption. The network settings of the motes in these models (transmission power and the distribution of messages to parents) are parameters that can be set to predict the quality properties for different configurations of the network.

4.4.2 Monitor

The monitor function has a dual role. On the one hand, it keeps track of the environment in which the managed system operates and reflects the relevant changes in the environment model. On the other hand, the monitor keeps track of changes of the managed system and updates the managed system model accordingly.

Figure 4.9 shows the basic workflow of the monitor function, which can be activated in different ways. The workflow can be triggered externally, for instance by a probe that finished a sensing cycle. The workflow may be triggered periodically, based on a predefined time window. Alternatively, the workflow may process sensor data in continuous cycles, i.e. the workflow cycle automatically restarts after the previous cycle has been finished.

After triggering, the monitor collects data taken from sensors. Depending on the domain at hand, different types of sensors can be used to track the environment and the managed system. For instance, the network infrastructure of DeltaIoT is equipped with sensors that track the signal-to-noise ratio (SNR) between source and destination motes. These values, which can be accessed via the management interface at the gateway, determine the level

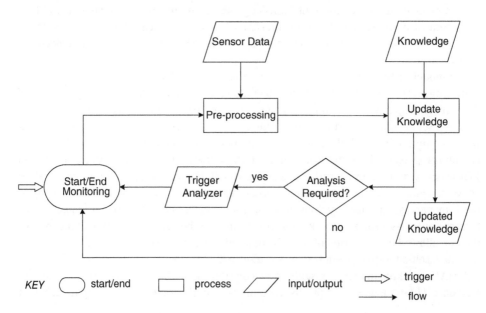

Figure 4.9   Basic workflow of the Monitor function.

of interference on the links coming from noise in the environment. On the other hand, the management interface provides access to various elements of the network, including the current settings of motes and statistical data on the quality properties of the network.

The newly collected data may require some form of pre-processing before it can be used to update the knowledge models. Pre-processing is a domain-specific step for which different mechanisms can be applied. Relatively simple mechanisms are filtering the newly collected data, for instance to remove noise, or aggregating the data, for example by determining the average and standard deviation of data values. More advanced techniques, such as Bayesian estimation or a deep neural network may be used to extract meaningful or timely knowledge from the raw data collected by the sensors. In DeltaIoT, for instance, a simple Bayesian learning mechanism may be used that uses the sequence of observations of SNR values of the communication links to update the values that represent the uncertainty of interference on the links of the network.

After the pre-processing step, the monitor uses the current knowledge and the pre-processed data to update the knowledge models. Such updates are domain-specific and can be as simple as changing the value of a simple data structure up to applying a complex graph transformation to represent an observed architectural reconfiguration of the managed system. In DeltaIoT, the actual values of the quality properties such as packet loss and latency may be represented as simple variables that can be updated. On the other hand, changes in the interference of certain links or traffic generated by motes will require an update of the corresponding elements of the relevant models of the knowledge.

When the knowledge models are updated, the monitor may perform a check to identify whether an analysis step is required or not. Such a check is another domain-specific task for which different techniques can be applied. A simple example is a technique that checks whether particular values of the knowledge have changed during the update and surpass particular thresholds. A more advanced technique may check for trends or patterns after the update of the knowledge. If such conditions are detected, the analyzer function will be triggered to continue the MAPE workflow. For example, in DeltaIoT, the analyzer may be triggered when the packet loss, latency, or energy consumption of the network has changed beyond given ranges.

4.4.3 Analyzer

The role of the analyzer function is to assess the up-to-date knowledge and determine whether the system satisfies the adaptation goals or not, and if not, to analyze possible configurations for adaptation.

Figure 4.10 shows the basic workflow of the analyzer function. Similar to the monitor, the workflow of the analyzer can be activated externally, for instance by the monitor, or it may be triggered periodically, based on a predefined time window, or the workflow may automatically restart after the previous cycle has been finished.

After triggering, the analyzer assesses the actual conditions based on the the current knowledge, in particular the adaptation goals, to determine whether adaptation needs to be initiated or not. Depending on the domain at hand, different mechanisms can be used to perform this assessment. A simple mechanism may check whether the system currently violates any of the individual adaptation goals and if so, an adaptation of the system is initiated.

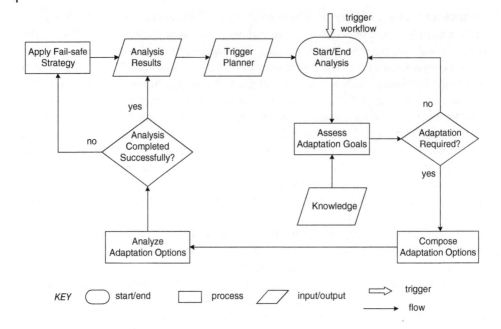

Figure 4.10 Basic workflow of the Analyzer function.

A more advanced mechanism may determine the utility of the system by combining the weighted values of the relevant quality properties. The decision to initiate adaptation or not may then be based on comparing the current utility with a threshold value. Alternatively, the change of utility over a time horizon may be used to make a decision. For instance, in DeltaIoT, the analyzer may simply compare the current values of packet loss and latency with the thresholds defined by the adaptation goals and initiate adaptation if any of these thresholds is violated.

When no adaptation is required, the workflow immediately ends. Otherwise, the analyzer composes the *adaptation options*. The adaptation options define a set of configurations that can be reached from the current configuration by adapting the managed system. The *adaptation space* denotes the set of adaptation options at any point in time. Different mechanisms can be applied to determine this set. A simple mechanism may combine all the possible parameter settings of the managed system that can be used to adapt the system. Such an approach would require that the parameters have discrete domains (or can be easily discretized) and that the adaptation space is limited in order to perform analysis within the available time window. Instead of exhaustively determining the adaptation space, a mechanism may start from the current configuration and determine adaptation options based on some predefined distance metric. For instance, only parameter settings with a maximum difference compared to the current settings are considered. This approach could be applied for systems where adaptation requires the system to move only slowly from one locality to another (in terms of differences in parameter settings). For complex managed systems, algorithms that directly determine adaptation options may no longer be applicable. Such systems may require advanced mechanisms such as data mining or machine learning to explore large adaptation spaces and identify suitable adaptation options. In general, the adaptation space may be static, change slowly over time, or be very dynamic. For DeltaIoT, each adaptation option corresponds to a different combination of settings of the motes of the network; the domain of the settings is determined based on all the values within the range of possible values for the transmission power and distribution of packets that can be used to adapt the system.

Once the adaptation space is determined, the adaptation options are analyzed. The aim of the analysis is to determine the "goodness" of each adaptation option. The goodness of an adaptation option is defined in relation to the adaptation goals and can be determined using a wide variety of mechanisms. Analysis mechanisms differ in the way they use knowledge over time: reactive mechanisms are primarily concerned with knowledge from the past, active mechanisms focus on present knowledge, the state of the system as it is now, while proactive mechanisms look into and predict the future ("what-if analysis"). The choice of a concrete mechanism will depend on the adaptation problem at hand. A simple mechanism may analytically compute an expected quality value for each adaptation option based on a table with current configuration settings. An advanced mechanism may apply model checking over models of the system to determine the expected accumulated utility over the horizon per adaptation option. For DeltaIoT, the analyzer may run simulations of up-to-date quality models of the system to determine the expected quality properties for each adaptation option in the next cycle.

When the analysis of the adaptation options completes successfully, the planner function will be triggered to prepare a plan for adapting the managed system based on the analysis results. However, analysis may not complete successfully for various reasons that depend on the domain at hand. One reason could be that none of the analyzed adaptation options comply with a minimum criterion to be considered for adaptation of the managed system. Another reason could be that the time window to perform the analysis of the adaptation options may be exceeded before any useful results are produced. In such cases, the analyzer will apply a fail-safe strategy before the planner is triggered. The concrete choice of a fail-safe strategy may depend on various conditions, such as the criticality of the system and its current situation, the reason why analysis was not completed successfully, and the partial analysis results that are available. Possible strategies could be to not adapt the system, to adapt the system to a predefined configuration, or to bring the system to a safe stop. For example, in DeltaIoT, default settings may be applied to the system (set the transmission power to maximum and duplicate the messages to all parents) in the event that the system violates the adaptation goals and no alternative configuration could be found within the time window for analysis.

4.4.4 Planner

The role of the planner function is to select the best adaptation option and generate a plan for adapting the managed system from its current configuration to the new configuration defined by the best adaptation option. In the event that a fail-safe strategy needs to be applied, the planner only needs to generate an adaptation plan to bring the system to a safe state.

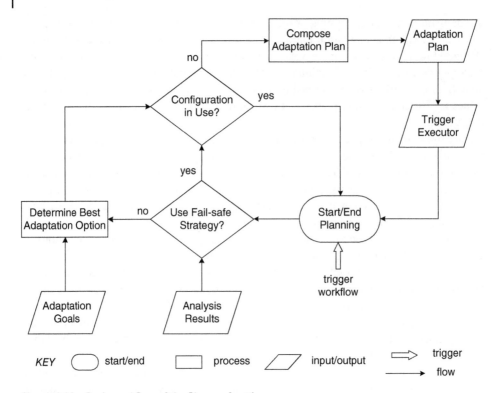

Figure 4.11 Basic workflow of the Planner function.

Figure 4.11 shows the basic workflow of the planner function. The workflow of the planner is usually activated externally, by the analyzer.

After triggering, the planner uses the analysis results to check if a fail-safe strategy needs to be applied. If that is not the case, the planner determines the best adaptation option based on the adaptation goals. Depending on the domain at hand, different mechanisms can be used to select the best adaptation option. We explain two possible mechanisms. If the adaptation goals are defined as a set of rules, the adaptation options can be ranked as follows. Assume the adaptation goals consist of a set of threshold goals (a threshold goal should keep a quality property of the system under a given value) and a goal that aims to minimize or maximize a quality property of the system. The selection of the best adaptation option starts by dividing the adaptation options in two groups: those that are compliant with the rules that define threshold values and those that are not compliant with these rules. Then the group of the options compliant with the threshold goals are ranked based on the analysis results for the quality property that needs to be minimized or maximized. The adaptation option at the top is then selected for adaptation. If the adaptation goal is defined as the maximized accumulated utility over the horizon, the adaptation options are simply ranked based on the expected utility determined during analysis. The option that is ranked best is then selected for adaptation. In DeltaIoT, the first selection mechanism described can be used since the adaptation goals are defined as threshold goals for packet loss and latency, and a goal that aims at minimizing energy consumption.

Once the best adaptation option has been determined, or if the fail-safe configuration needs to be applied, the planner checks whether the new configuration is currently in use. If this is the case, the planner function ends. Otherwise, the planner composes a plan to adapt the managed system. A plan defines the appropriate course of action, i.e. the actions that need to be performed to adapt the managed system from its current configuration to the new configuration. An *adaptation action* defines a unit of activity that needs to be exercised on the managed system in order to adapt it. In general, an adaptation action can define any sequence of instructions that the managed system understands and that can be executed to adapt it. What constitutes an adaptation action is domain-specific and needs to be defined for the concrete adaptation problem. For instance, in DeltaIoT, an adaptation action is defined as the set of new settings of a mote that need to be exercised to adapt the network, i.e. the changes of transmission power and distribution of packets of a mote for the links to its parents.

In general, the choice for a planning mechanism is determined by the (adaptation) problem at hand. Specifically for self-adaptive systems, the quality and timeliness of the plan are two important criteria for selecting a planning mechanism. Quality relates to the likelihood that a generated plan will eventually meet the adaptation goals. Quality of planning is particularly important for domains with strict goals. Timeliness relates to the time required to generate a plan. Timeliness of planning is important for domains with time constraints between required adaptations. Clearly, quality and timeliness are conflicting criteria, hence a tradeoff needs to be made.

Over time, numerous planning mechanisms have been defined, in particular by the Artificial Intelligence (AI) community. Reactive planners are usually fast and may exploit reactive plans, for instance in the form of condition-action rules or finite state machines. These plans are typically parameterized and can be instantiated for the concrete setting. Other planning approaches rely on languages for representing planning problems that are based on state variables. Applied to self-adaptive systems, each possible state is an assignment of values to the state of the managed system, and an action determines how the values of the state variables change when that action is taken. Generating a plan then boils down to finding the sequence of actions that bring the managed system from its current state to the new state after adaptation. Similarly to many other computational approaches, this mechanism may suffer from a combinatorial explosion of the state space. For DeltaIoT, a simple dedicated planner can be used that determines the adaptation steps that need to performed to adapt the network settings mote by mote and link by link from the current to the adapted configuration.

Once the adaptation plan has been generated, the planner triggers the executor function.

4.4.5 Executor

The role of the executor function is to execute the adaptation plan, adapting the managed system from its current configuration to the new configuration.

Figure 4.12 shows the basic workflow of the executor function. The workflow of the executor is usually activated externally, by the planner.

When the executor is triggered, it selects an adaptation action of the adaptation plan and enacts this action on the managed system via an effector. If the plan is defined as

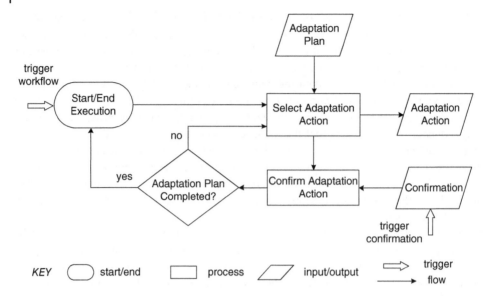

Figure 4.12 Basic workflow of the Executor function.

an ordered sequence of adaptation actions, the first action of the sequence is selected for execution. Without any particular ordering, any adaptation action can be selected. For DeltaIoT, for instance, the adaptation actions may be ordered taking into account the time-synchronization schedule of the underlying network.

In general, the execution of adaptation actions may take some time to produce their effect. For instance, sending a message with instructions to change the settings of a mote in DeltaIoT will take some time as it needs to respect the time schedule of the underlying network. To that end, the basic executor waits for a confirmation when it exercises an adaptation action. Such confirmation may reach the executor in different ways. One approach is to let the managed system notify the executor, directly or indirectly, via the effector. Another approach is to delegate the task to the monitor function; however, this will require that the monitor is aware of the adaptation actions and how their effects can be observed. Yet another approach could be to let the executor wait for predefined time windows after it has executed adaptation actions. Which approach is used will depend on the concrete setting at hand.

When the executor receives a confirmation of an adaptation action, it will check whether the adaptation plan is completed or not. If not, another adaptation action will be selected and exercised on the managed system. As soon as the adaptation plan is completely executed, the executor function ends. The executor will then wait for the next trigger of the planner function to execute the next adaptation plan.

4.5 Software Evolution and Self-Adaptation

We have explained how the manageability problems of complex software systems were the primary motivations for self-adaptation. However, traditionally, handling change in software-intensive systems has been realized through software evolution. This raises the question of how software evolution and self-adaptation relate to one another. Before we elaborate on this relationship, we start with a brief explanation of software evolution management and self-adaptation management.

4.5.1 Software Evolution Management

The term *software evolution* refers to the process of repeatedly updating software after its initial deployment. The reasons why software needs to be updated are very diverse. Updates may be required to correct discovered problems; to anticipate potential faults; to deal with changes in the environment, such as updates of the underlying platform or changes in network technology; to improve the performance or utility of the system; or to add new functionality to the system. With the increasing demand for business continuity (i.e. system services need to be available 24/7), there is a growing need to update software systems in the field, with minimal or no interruption of their services. Agile practice, which emerged around the same time as autonomic computing, follows an evolutionary software development process, encouraging iterative and incremental development of software in short cycles allowing rapid and flexible response to change.

Figure 4.13 shows the basic artifacts and activities involved in evolution management of a software system.

Evolution management is performed by humans, supported by tools. Different stakeholders may be involved, including managers, architects, developers, testers, and users. A wide variety of tools can be used, including design tools, simulators, tools for testing and verification, build tools, etc. Evolution management is centered around two main types of system artifacts: development models and the system implementation. Development models can be very diverse, including specifications of requirements, design, deployment, processes, etc. The system implementation includes the actual code, supporting infrastructure, deployment scripts, and other related artifacts.

Evolution management can be triggered externally, for instance based on a request for new functionality, or as part of a strategic release plan for the software. Evolution can also be triggered internally, based on the observation of the running system, for example when bugs are detected or the system fails to perform as required.

The basic activities of evolution management are to prepare the system update and to enact the evolution of the system. The preparation of a system update requires stakeholders to analyze the request for change, plan the update of the system, and evolve and test the system accordingly. The update is then enacted on the running system. Typical updates include changing components, adding new components, integrating platform updates, etc. Performing such updates in the field requires appropriate infrastructure. The needs for such infrastructure depend on the characteristics of the system (for instance stateless versus stateful) and the types of updates that need to be applied. Enactment typically starts with loading the updated software, adding or replacing parts of the running system by changing bindings between system elements, and activating the newly deployed parts of the system.

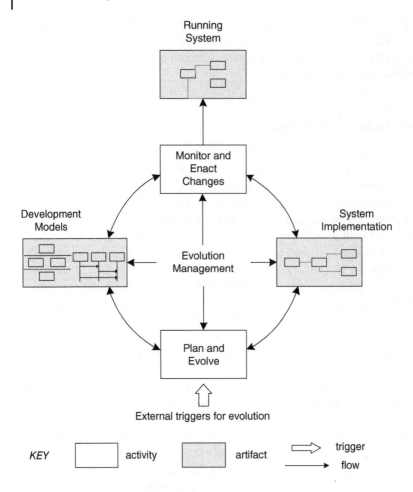

Figure 4.13 Basic artifacts and activities of evolution management.

4.5.2 Self-Adaptation Management

We direct our attention now to self-adaptation management of a software system. Figure 4.14 shows the basic activities and artifacts involved in self-adaptation management.

Self-adaptation management refers to a process of system adaptation that is performed automatically, possibly supported by human stakeholders. Self-adaptation management is centered around two types of system artifacts: runtime models (aka knowledge) and the running system (aka the managed system). These artifacts are used by four basic activities – monitor, analyze, plan, and execute – that together form a feedback loop. This feedback loop monitors the running system and its environment and adapts the running system when needed to realize the adaptation goals. Such adaptation may be triggered by a violation of one or more adaptation goals; alternatively, self-adaptation may pro-actively anticipate such violations. Self-adaptation requires that the running system is equipped with infrastructure to monitor the system and enact changes in a consistent manner.

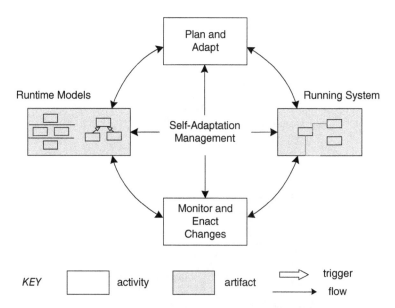

Runtime Models

Running System

Plan and Adapt

Self-Adaptation Management

Monitor and Enact Changes

KEY activity artifact trigger flow

Figure 4.14 Basic artifacts and activities of self-adaptation management.

4.5.3 Integrating Software Evolution and Self-Adaptation

Let us now direct our attention to the integration of evolution management and self-adaptation management. Figure 4.15 shows how evolution management and self-adaptation management integrate with one another and how the activities of the two cycles work together.

Evolution management and self-adaptation management are complementary activities that deal with different types of change. Evolution management deals with *unanticipated change* and requires human involvement. Self-adaptation management on the other hand deals with *anticipated change* and may or may not involve humans in the loop. By "dealing with anticipated change," we mean that the system has been built such that it can detect the change and handle it in some way.

The activities of evolution management and self-adaptation management interact with one another in different ways. Two principal classes of such interactions are shown in Figure 4.16.

The left part of Figure 4.16 shows a scenario where adaptation management triggers evolution management. In this particular case, the analysis element of the feedback loop discovers a problem for which no mitigation plan is available. This triggers evolution management, which will process the request. The evolved planner models and corresponding implementations will be added. Finally, the running system will be updated, resolving the initial problem.

The right part of Figure 4.16 show a scenario where evolution management triggers adaptation management. In this specific case, an update of the running managed system is requested. This request will trigger evolution management, which will initiate an update of the system model and a corresponding evolution of the system implementation. After that, adaptation management will be triggered to update the runtime model of the system and

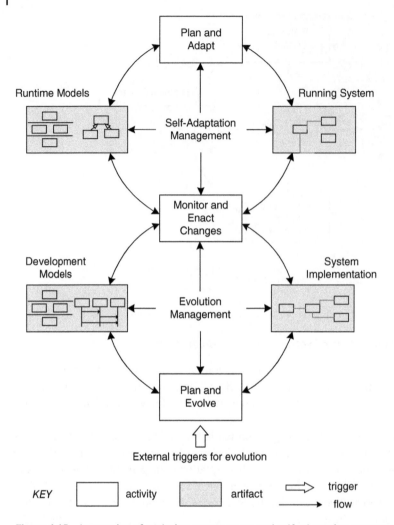

Figure 4.15 Integration of evolution management and self-adaptation management.

the running code accordingly. An advanced scenario would be one where the system goals evolve, which requires not only an evolution of both the managed system and the feedback loop, but also a complex synchronization schema to organize the updates of the different artifacts in a safe and consistent manner.

4.6 Summary

Autonomic computing aims to free system administrators from system installation, operation, and maintenance by letting computing systems manage themselves based on high-level objectives. The capabilities of an autonomic computing system rely on a feedback loop.

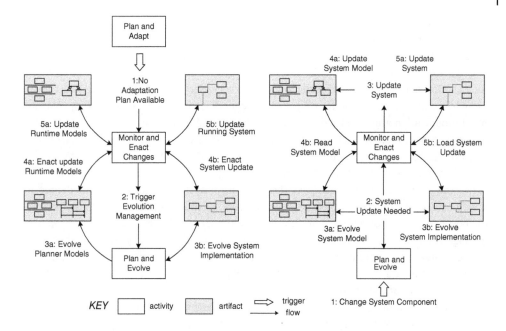

Figure 4.16 Two principal classes of interaction between evolution management and self-adaptation management. Left: adaptation management triggers evolution management; Right: vice versa.

Four essential maintenance tasks can be distinguished: self-optimization, self-healing, self-protection, and self-configuration.[4]

A self-optimizing system continuously seeks opportunities to optimize its resource usage while maintaining its required quality objectives. An example is an IoT system that optimizes the energy consumed by its motes while keeping the packet loss and latency below given thresholds.

Self-healing is the capability of a system to detect and recover from failures in order to deliver the required quality objectives or degrade gracefully. An example is an IoT system that continuously tracks failing motes, diagnoses the problem, and recovers itself to maintain its quality of service.

A self-protecting system defends against malicious attacks that may jeopardize the required quality objectives. An example is a time-synchronized IoT system that continuously scans the network traffic to anticipate and defend against jamming attacks by adapting its time-synchronization schedule to preserve its quality objectives.

Self-configuration is the capability of a system to automatically install and configure new elements without interrupting the managed system's operation. An example is the automated integration and configuration of new elements in an IoT network, while the rest of the network adapts seamlessly.

---

4 These four tasks were initially proposed by IBM and then confirmed by many others. However, in general, there is a huge variety of related tasks that are often referred to as "self-*".

The reference model of a managing system comprises four primary elements, which share knowledge: monitor, analyzer, planner, and executor. These elements, commonly referred to as MAPE-K, realize the basic functions of any self-adaptive system.

The knowledge function provides the knowledge and mechanisms that are required for the MAPE functions to work together and realize self-adaptation. In a concrete realization, knowledge can be mapped to four types of basic models: managed system model, environment model, adaptation goal model, and MAPE working models.

The monitor function tracks relevant aspects of the environment and the managed system and keeps the corresponding models up-to-date. The analyzer function assesses the up-to-date knowledge to determine whether the adaptation goals are satisfied or not. If this is not the case, the analyzer analyzes the adaptation options, determining the degree to which each of the options is expected to achieve the adaptation goals. The planner selects the best adaptation option based on the adaptation goals and generates a plan to adapt the managed system accordingly. If for some reason no valid adaptation option could be found by the analyzer, the planner will generate a plan to bring the managed system to a fail-safe state. Finally, the executor executes the adaptation plan, adapting the managed system from its current configuration to the new configuration.

Software evolution management refers to the process of repeatedly updating software after its initial deployment. Evolution is a human-driven process, supported by tools. Self-adaptation management, on the other hand, refers to the online adaptation of the system and is a machine-driven process that may be supported by humans.

Evolution management and adaptation management are complementary processes that can interact in two directions. Evolution management can support adaptation management when the managing system encounters problems that were not fully anticipated. Adaptation management can support evolution management to keep the running system and the corresponding runtime models aligned when the system or its goals evolve.

Table 4.1 summarizes the key insights of Wave I.

Table 4.1   Key insights of Wave I: Automating Tasks.

• Automating tasks is a key driver for self-adaptation. This driver originates from the difficulty of managing the complexity of large-scale interconnected computing systems.

• Four essential types of maintenance tasks for automation are self-optimization, self-healing, and self-protection, and self-configuration.

• Monitor, Analyse, Plan, Execute + Knowledge, MAPE-K in short, provides a reference model for a managing system. The MAPE-K functions are intuitive; however, their concrete realization introduces significant scientific and engineering challenges.

• Evolution management deals with unanticipated change, which requires human involvement. Self-adaptation management deals with anticipated change, which may be supported by humans.

• Evolution management and self-adaptation management are complementary activities that can work in tandem to support one another.

4.7   Exercises

4.1   Utility function for self-healing: level H

Consider a self-healing scenario of a simplified IoT network, similar to the one described in this chapter. Perform a utility calculation in an unperturbed system and identify the optimal system configuration. Then, introduce a scenario where a mote fails, re-run the optimization, and show that it produces a different outcome. Interpret and discuss your results.

4.2   MAPE functions for self-optimization in DeltaIoT: level D

Setting. Consider the self-optimization scenario of DeltaIoT shown in Figure 4.3. Assume the system has three adaptation goals: packet loss $\leq$ 10 %, latency $\leq$ 5% of the cycle time, and minimize energy consumption.

Task. Specify the necessary knowledge models that enable a managing system to seek opportunities to optimize the energy consumption in this scenario. Instantiate the workflows for the different MAPE functions. Explain how the different MAPE functions share the knowledge models and collaboratively achieve the self-optimization objective.

Additional material. For more information about DeltaIoT, see the DeltaIoT website [213].

4.3   Implementation of self-optimization in DeltaIoT: level D

Implement your design for the self-optimization scenario using the DeltaIoT simulator (see Exercise 4.2). Define a number of concrete settings, and compare your solution with a non-adaptive approach. Assess your results and discuss potential improvements of your solution. The DeltaIoT simulator can be downloaded from the DeltaIoT website [213].

4.4   Implementation of self-adaptation for other tasks in DeltaIoT: level D per task

Repeat Exercise 4.3 for the other essential maintenance task scenarios, i.e. self-healing shown in Figure 4.4, self-protection shown in Figure 4.5, and self-configuration shown in Figure 4.6.

4.5   Implementation of self-adaptation with utility function in DeltaIoT: level D

Setting. Consider the TAS artifact described in Exercise 1.4. Design a MAPE-K feedback loop that deals with service failures. Consider again two adaptation goals – service failures and cost – and define a utility function to determine the adaptation strategy.

Task. Implement your design ensuring that the weights of the utility function for both goals are implemented as variables. Test your solution with different values of the weights. Evaluate the impact of the different settings. Assess the results and make a tradeoff analysis between the two solutions.

Additional material. The latest version of TAS can be downloaded from the TAS website [212]. For background information about TAS, see [200].

4.6 Self-adaptation to mitigate denial of service attacks in Hogna: level W

Setting. The Hogna artifact offers a platform to evaluate self-management strategies in a Cloud environment. The artifact comes with a sample application that can be automatically deployed and managed in the Amazon EC2 Cloud. In addition, a workload generator is available, which can be configured together with scripts to visualize system data. The platform offers a basic self-management strategy that relies on simple scaling rules to trigger the addition or removal of virtual machines based on changing workload.

Task. Download the Hogna artifact and install it. Run the sample application and experiment with different workloads. Adjust the threshold values of the rules used in the basic adaptation strategy and observe the effects. Now consider the possibility of a denial of service attack. Such an attack floods the Cloud environment with a high volume of traffic, preventing access for legitimate users. Enhance the basic self-management strategy with support to handle denial of service attacks. Emulate a denial of service attack and test the management strategy. Critically assess your solution.

Additional material. The Hogna artifact is available for download at [18]. More information about the artifact can be found in [17]. For further information of denial of service attacks, see for instance [140, 160].

4.8 Bibliographic Notes

IBM's original manifesto and a white paper, which were both published around 2000, introduced the notion of autonomic computing, a deliberately chosen term with a biological connotations [99, 100]. The company puts forward autonomic computing as the inevitable answer to the manageability problems that were caused by the exponential growth in complexity of software applications and the environments in which they operated.

P. Oreizy et al. pointed to the need for software systems to retain full plasticity throughout their lifetime, and stated that self-adaptation will provide the key to this [151]. Around the same time, J. Magee and J. Kramer emphasized the need for dynamic software architectures in which the organization of components and connectors can dynamically change during system execution [131].

D. Garlan and B. Schmerl contrasted traditional mechanisms to detect and recover from errors, which are typically wired into applications at the level of code, with externalized adaptation mechanisms that maintain a set of models at runtime external to the application as a basis for identifying problems and resolving them [80].

S. Russell and P. Norvig referred to utility for an action or a choice as "the quality of being useful" [169]. T. Glazier et al. elaborated on the application of utility functions in self-adaptive systems [88]. G. Tesauro and J. Kephart demonstrated how utility functions enable the continual optimization of the use of computational resources in a dynamic, heterogeneous environment [186]. J. Kephart and W. Walsh analyzed the connections among rules, goals, and utility functions [113].

In their seminal work, J. Kephart and D. Chess characterized the essential maintenance tasks for adaptation [112]: self-configuration, self-optimization, self-healing, and self-protection. The authors introduced the MAPE-K model, which defines the primary functions of self-adaptation. D. Ghosh et al. surveyed self-healing systems [87], and E. Yuan et al. presented the results of a systematic literature review on self-protecting software [217]. N. Khakpour et al. presented an interesting work on self-protection to deal with additional threats that are introduced through the adaptation process itself [114]. M. Salehie and L. Tahvildari highlighted different types of knowledge models [170], which were later systematically integrated into the FORMS reference model [206].

S. Kounev et al. provided an interesting reference to what self-awareness is [119], A. Dey elaborated on context-awareness [61], and P. Sawyer et al. discussed requirements-aware systems [171].

K. Bennett and V. Rajlich described a research roadmap on software maintenance and evolution, focusing on improving speed and accuracy of change while reducing costs [28]. Fourteen years later, the second author emphasized that successful software requires constant change, which has led to evolutionary software development [161]. P. Oreizy discussed the relationship between software evolution and software adaptation and integrated both activities in a comprehensive and visionary adaptation methodology [151].

The Hogna artifact was developed at York University, Toronto, Canada by C. Barna et al. [17].

5

Wave II: Architecture-based Adaptation

The second wave directs the focus beyond the primary motivation for self-adaptation to the foundations of engineering self-adaptive systems. The first wave identified a set of principles and concepts, such as the essential maintenance tasks and the MAPE-K model. However, these principles and concepts do not provide an integrated perspective on how to engineer self-adaptive systems. In the second wave, the focus shifts to basic abstractions for self-adaptive systems from two complementary angles. On the one hand, the focus is on design abstractions that enable designers to *define* self-adaptive systems – i.e. these abstractions are essential for designers to structure self-adaptive systems and allocate the primary concerns of self-adaptation. On the other hand, the focus is on modeling abstractions that enable the system to *reason* about change – i.e. the abstractions offer an appropriate perspective on the managed system and its environment enabling the managing system to make effective adaptation decisions.

Already in the 1990s, different research groups in the USA and Europe raised concerns about the economical development of software that satisfies its objectives. The researchers pointed to the growing problems of the increasing scale of computing systems and the continuous change these systems face. To tackle these problems, new architectural styles and supporting design languages were invented that aimed at bringing the required flexibility for software systems to be easily modifiable, to enable both incremental development and the changing of systems in the field.

The architectural style called Weaves, for instance, which was developed by the Aerospace Corporation in the USA, defined a system as interconnected networks of components that consume data objects as input and produce data objects as output. Weaves networks can be automatically composed and recomposed based on data analysis performed by tools. Another example is the Darwin language, which was developed at Imperial College London with the objective of supporting the specification and construction of dynamically reconfigurable distributed systems. Darwin is a declarative language that supports static and dynamic component-based structures, which may evolve during execution. For instance, components can be instantiated and bound dynamically.

This line of pioneering work laid the basis for the second wave: Architecture-based Adaptation. Architecture-based Adaptation provides an answer to the need for a systematic engineering perspective to automate management tasks, which was the driver for the second wave. The emphasis is on two complementary aspects: on the one hand, the wave advocates an architectural approach for *defining* self-adaptive systems; on the other hand,

An Introduction to Self-Adaptive Systems: A Contemporary Software Engineering Perspective,
First Edition. Danny Weyns.
© 2021 John Wiley & Sons Ltd. Published 2021 by John Wiley & Sons Ltd.

the wave promotes the use of an architectural model for *reasoning* about self-adaptation during operation.

In this chapter, we elaborate on the motivation for an architectural perspective to self-adaptation. Then, we zoom in on a three-layer model that defines the basic activities of a self-adaptive system and allocates the essential concerns of self-adaptation to the different activities. Next, we elaborate on the use of an architectural model that provides the appropriate perspective on the managed system and its environment, enabling a managing system to reason about self-adaptation. We illustrate how an architectural model is applied in an example of a self-adaptive Web-based client-server system. Finally, we present a comprehensive reference model for self-adaptation that integrates the different architectural aspects of self-adaptive systems, i.e. reflection, MAPE-K and distribution.

LEARNING OUTCOMES

- To motivate and illustrate the rationale for an architectural perspective on self-adaptation.
- To explain the three-layer model for self-adaptive systems.
- To map the layers of the three-layer model to a concrete self-adaptive system.
- To explain the use of an architectural model to support self-adaptation during operation.
- To illustrate how an architectural model supports self-adaptation during operation.
- To explain the different perspectives of the encompassing reference model for self-adaptation.
- To apply the different perspectives of the reference model to a concrete self-adaptive system.

5.1 Rationale for an Architectural Perspective

Handling change can be realized in different ways. Traditionally, so called "internal mechanisms," such as exceptions (as a feature of a programming language) and fault-tolerant protocols are used to deal with change events. The concrete application of such mechanisms is often domain-specific and tightly bounded to the code. This makes it costly to build, modify, and reuse solutions based on internal mechanisms. In contrast, self-adaptation enables change to be handled by using "external mechanisms." In this approach, adaptation becomes the responsibility of elements external to the system that manage the system. External mechanisms have the advantage that they localize the adaptation concerns in separable system elements that can be analyzed, modified, and reused across different self-adaptive systems. External mechanisms also allow adding self-adaptation capabilities to legacy systems, even to systems for which the code may not be available. In addition, the use of external mechanisms paves the way for providing managing systems with an integrated perspective on the managed system in order to handle changing conditions.

Self-adaptation based on an external feedback loop was not the first external mechanism that was proposed to monitor, reason about, and adapt an underlying software system. One prominent example of a predecessor is computational reflection. Whereas traditional computation is computation about the problem domain, reflective computation is computation about the computation of the domain. Examples are a computation that tracks the performance of a system or a computation that activates some functionality when needed, for instance a window for a user in a particular context. Hence, a reflective computation does not directly contribute to realizing the domain concerns of a computing system. Instead, it contributes to the internal organization of the computing system and allows observing and modifying a program's execution at runtime. Most programming languages today support reflective computation though language primitives, libraries, and supporting executing environments. Some issues have been raised with the use of reflection in production code, such as increased complexity of debugging and performance overhead.

Another example of a predecessor of self-adaptation is a meta-object protocol that defines the rights and obligations to access, reason about, and manipulate the structure and behavior of systems of objects. A meta-object protocol allows, for instance, to create a new class, change the definition of a class, or delete a class. Meta-object protocols are nowadays often applied in design and development tools where such protocols are used to represent and manipulate the design artifacts. A meta-object protocol is one way to implement aspect-oriented programming, which is another external mechanism that allows adding additional behavior to existing code (in the form of a so-called advice) without modifying the code itself. Several criticisms were raised of aspect-oriented programming: most notably the fact that the control flow in an aspect-oriented application is obscured, but also regarding the unavoidable need for global knowledge of the system in order to reason about the dynamic execution of local aspects.

This brings us to the question of why a software architecture perspective to self-adaptation with an external feedback loop would be a suitable approach to deal with change at runtime. Numerous researchers and engineers have elaborated on this issue. We highlight a number of key arguments that have been put forward in favor of an architectural perspective on self-adaptation.

- *Separation of concerns*: Separation of concerns is a design principle that divides a software system into parts and allocates a distinct set of concerns that are of interest to stakeholders to each part of the system. A software architecture perspective on self-adaptation supports the separation of the domain concerns that relate to the functionality and goals of users of the system and are allocated to the managed system, and the adaptation concerns that relate to the adaptation goals and are allocated to the managing system. This separation of concerns results in higher flexibility in the design, maintenance, and reuse of the main parts of a self-adaptive system.
- *Integrated approach*: Specifications at the software architectural level span the full range of activities involved in engineering software systems, ranging from configuration, testing, and deployment to reconfiguration, and maintenance. As such, software architecture has the potential to offer an integrated approach for the development, operation, and evolution of self-adaptive systems.
- *Leveraging consolidated efforts*: Software architecture is centered on generic concepts and principles that are transferable to a wide range of application domains. Hence, architecture enables software engineers to build on consolidated efforts, such as specification

languages and notations, patterns and tactics, and formally founded architecture-based analysis and reasoning techniques.

- *Abstraction to manage system change*: Software architecture provides an appropriate level of abstraction to describe the dynamic changes of a system. Using a component-based model of the system, for instance, enables specification and reasoning about adaptation in terms of adding, removing, and binding self-contained elements. Abstraction allows us to focus on those facets of the system that are relevant to the concerns of adaptation and omit the rest.
- *Dealing with system-wide concerns*: Centering self-adaptation on an architectural model of the system offers a global perspective of the system and enables monitoring and reasoning about important system-level properties and integrity constraints. The main concerns of software systems that are the subject of adaptation, such as performance, reliability, and security, depend to a large extent on the software architecture of the system.
- *Facilitating scalability*: Software architecture supports composition, including hierarchical composition, which is particularly useful to specify and reason about a system at different levels of granularity. Composition facilitates the self-adaptation of large-scale complex applications.

The first three arguments motivate the need for an architectural perspective from the point of view of defining self-adaptive systems. The last three arguments put the emphasis on the use of architectural models for reasoning about adaptation during operation. Together, these arguments have paved the way for researchers and software engineers to apply the architectural principles of abstraction and separation of concerns as a foundation for engineering self-adaptive systems.

5.2   Three-Layer Model for Self-Adaptive Systems

The three-layer model for self-adaptation offers an architectural perspective on defining self-adaptive systems. The model distinguishes the main high-level activities involved in self-adaptation and allocates concerns to these activities.

The three-layer model is inspired by the classic architecture of robotic systems, which consists of three layers. The basic layer comprises simple reactive behaviors that allow the robot to perform basic tasks based on sensor input, for instance moving along a path and avoiding obstacles. The middle layer comprises reactive plans that sequence and monitor the basic tasks of the robot. A reactive plan may, for instance, instruct a robot to move to a location, pick up a packet, move to a destination, and drop the packet. Finally, the top layer decides how to achieve high-level goals. This layer uses reasoning mechanisms that typically take into account long term information to decide on the strategic course of action of the robot over time.

Figure 5.1 shows the simple yet powerful three-layer model for self-adaptive systems that is based on the classic architecture of robotic systems. The model separates the main concerns of self-adaptation, which are mapped to the different layers. The complexity of the activities increases from bottom to top. This requires each layer to operate at a specific time scale. A layer higher in the hierarchy typically operates substantially slower than the layer below.

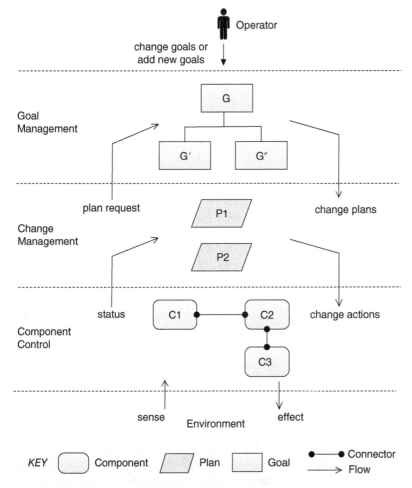

Figure 5.1 Three-layer model for self-adaptive systems.

5.2.1 Component Control

The bottom layer, *Component Control*, consists of the interconnected components of the system. These components sense and effect the environment in order to realize the goals for the users of the system. The component control layer may contain internal mechanisms to adjust the system behavior. For instance, exceptions may be used to deal with events that may disrupt the normal flow of execution of the system. To realize self-adaptation, component control needs to be instrumented with mechanisms to report the current status of the system to higher layers as well as mechanisms to support runtime modification, such as component addition, deletion, reconnection, and reconfiguration.

5.2.2 Change Management

The middle layer, *Change Management*, which realizes a MAPE-based workflow, is centered on a set of plans, which are typically predefined. These plans can be enacted by change management to adapt the component control layer. In particular, change management reacts in response to status changes of the bottom layer by analyzing the changes, selecting a plan, and executing this plan through change actions that adapt the configuration of the bottom layer. Change actions include adjusting operation parameters of components, removing failed components, adding new components, and changing interconnections between components. If a condition is reported that cannot be handled by the available plans, the change management layer invokes the services of the goal management layer.

5.2.3   Goal Management

The top layer, *Goal Management*, which also realizes a MAPE-based workflow, is centered on a set of high-level goals. This layer can produce plans for change management in response to requests for plans from the layer beneath. Such a request will require goal management to reason about the feasibility of achieving the goals in the changed setting and, if necessary, adapting the means for achievement or changing the goals. Concretely, a request from change management will trigger goal management to analyze the situation, select alternative goals based on the current status of the system, and generate plans to achieve these alternative goals. The new plans are then delegated to the change management layer. Goal management can also be triggered by stakeholders that want to change goals or introduce new goals. New goals may be accompanied by plans to realize the goals, or new plans may be synthesized automatically. Automatic synthesis of plans for new goals is often a complex and time-consuming task.

5.2.4   Three-Layer Model Applied to DeltaIoT

Figure 5.2 instantiates the three-layer model for DeltaIoT. We start by illustrating component control and change management in general. Then we elaborate on goal management using a concrete scenario when a mote fails, which is illustrated through the hierarchy in the figure.

In DeltaIoT, the component control layer consists of the network of motes that sense the environment and the gateway that collects the sensor data. The gateway is connected to a front-end component that provides the building security monitoring services to the users. Component control may implement internal mechanisms to adjust the system behavior when needed. For instance, the network infrastructure may have a built-in error checking function that requests a retransmission of packets if the received data is corrupted. The DeltaIoT network is equipped with a management interface that is deployed at the gateway. This interface enables the collection of operating network data and adapting the settings of the motes.

Change management in DeltaIoT consists of parameterized plans that can be used to adapt the network settings. Parameters of plans are the identifiers of a mote and its parents and variables for the transmission power and distribution of packets along the communication links to parents. Change management uses the management interface to track changes in the network. When the layer observes a violation (or potential violation) of the packet loss or latency of the network, it analyses the situation, determines a configuration

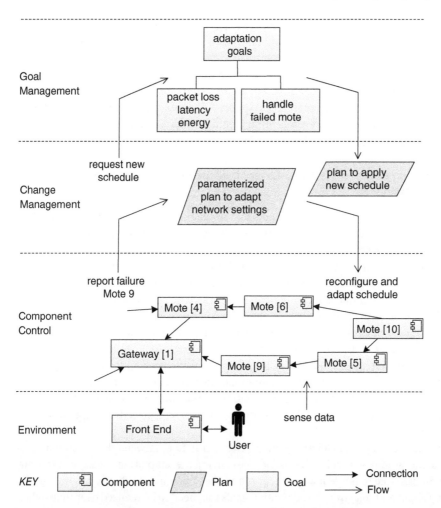

Goal
Management

Change
Management

Component
Control

Environment

Figure 5.2 Three-layer model applied to DeltaIoT with a scenario that illustrates an automatic reconfiguration of the network when a mote fails.

to adapt the network, and instantiates a plan for each mote that needs to be adapted. A plan is instantiated by selecting a mote that requires adaptation, assigning the identifiers of this mote and its parents, and assigning values for the new settings of the transmission power and the distribution of packets to the corresponding parameters. The set of instantiated plans, one for each mote that needs to be adapted, is then enacted on the network through the management interface.

In DeltaIoT, goal management can be added as an additional layer on top of change management. The scenario in Figure 5.2 illustrates how goal management may deal with a failing mote (automatically, without intervention of stakeholders). When component control detects that Mote [9] has failed, it informs change management. Since change management has currently only one plan available to deal with (potential) violations of the packet loss or latency goals of the network, it cannot handle the failure of the mote. Hence change management requests a plan from the goal management layer to handle the problem. Goal management will search for a corresponding goal ("handle failed mote") and will instantiate a plan that allows the failure of the mote to be handled. This plan is then sent to change management, which will execute the plan on the component control layer – i.e. it will reconfigure the links of the network between the motes and adapt the schedule for the new network configuration (as shown in Figure 4.4). Once the adaptation is completed, the change management layer can take up its regular tasks again and adapt the network settings when needed to achieve the adaptation goals.

5.2.5   Mapping Between the Three-Layer Model and the Conceptual Model for Self-Adaptation

Let us now look at how the three-layer model for self-adaptive systems that follows the basic structure of the architecture of robotic systems (Figure 5.1) relates to the conceptual model of a self-adaptive system (Figure 1.2).

The component control layer, which deals with the concerns of the users of the system, directly maps to the managed system, which is a basic element of the conceptual model. Hence, the set of interconnected components in the three-layer model realize the managed system. This basic level needs to provide probes and effectors to the level above.

The two other layers of the three-layer model, change management and goal management, map together to the managing system in the conceptual model. The adaptation concerns, for which the managing system is responsible, are split into two parts in the three-layer model. In particular, change management is concerned with the actual execution of the plans that realize the adaptation of the component control layer, while goal management is concerned with determining the concrete goals that adaptation needs to realize over time.

Hence, the three-layer model makes an explicit distinction between the functionality of the managing system, which is responsible for realizing the adaptation goals given the current conditions – i.e. change management – and the functionality that is responsible for reasoning about and selecting the adaptation goals that need to be realized over time, when the conditions change – i.e. goal management. The principal criteria for placing these two responsibilities in different layers are the complexity of the activities of the different layers to realize these responsibilities and the different timescales at which the activities work. Change management acts in a reactive manner through feedback actions in response to changes of the system in the current context. Goal management, on the other hand, acts in a deliberative manner through reasoning about the changing conditions of the system and its context over time and adapting the goals for adaptation accordingly.

Consequently, separating the realization of the current adaptation goals and the strategic selection of the adaptation goals over time, and treating them as first-class citizens, is a fundamental architectural principle for the engineering of self-adaptive systems.

5.3   Reasoning about Adaptation using an Architectural Model

We move our focus now from the offline definition of self-adaptive systems to the online reasoning of the system about adaptation using an architectural model. We start by outlining a general runtime architecture for an architecture-based approach to realizing self-adaptation, focusing on the use of an architectural model. Then we zoom in on a concrete realization of the approach using an example of a Web-based client-server system. The principles of architecture-based adaptation presented in this chapter are based on the Rainbow framework, which is a pioneering realization of architecture-based adaptation.

5.3.1   Runtime Architecture of Architecture-based Adaptation

Self-adaptation based on an external feedback loop requires an appropriate representation of the system and its context that allows the system to reason about its dynamic behavior and make adaptation decisions. An architectural model represents the system as a composition of components, their interconnections, and their properties of interest. An architectural model can provide a global perspective of the system and expose important system-level properties and integrity constraints.

Figure 5.3 shows the runtime architecture of an architecture-based approach equipped with an architectural model to realize self-adaptation.

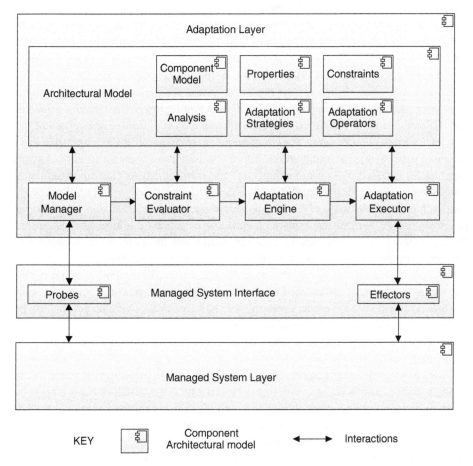

Figure 5.3   Runtime architecture of an architecture-based adaptation approach.

The managed system layer provides the domain functionality to users. This layer offers an access interface with probes to observe and measure various states of the system and its environment. The probes can publish the system information or the information can be queried from the probes. The interface also offers access to effectors that enable modifications of the managed system to be carried out, such as adjusting system parameters or adding and removing elements from the system configuration.

The adaptation layer comprises a set of components that form a feedback loop. These components share an architectural model of the system. The feedback loop makes adaptation decisions realizing a single level of adaptation.

The four components of the feedback loop implement the four basic functions of a managing system: monitor, analyze, plan, and execute. In brief, the model manager tracks the state of the underlying managed system via the probes to maintain an architectural model of the system. The constraint evaluator periodically checks a set of constraints over the architectural model and triggers the adaptation engine if a constraint violation is detected. The adaptation engine then determines the course of action that is required to handle the violations and the adaptation executor carries out the necessary adaptation actions to the running system via the effectors.

Central to the operation of the feedback loop components is the architectural model, which encodes domain-specific knowledge of the system. The architectural model, which is shared by the feedback loop components, comprises six elements: component model, properties, constraints, analyses, adaptation strategies, and adaptation operators.

The *component model* represents the managed system as a graph of interacting computational elements. Nodes correspond to components and represent the system's principal computational elements and data stores. Arcs correspond to connectors and represent the pathways for interaction between the components. A component in the graph may itself represent a subsystem, enabling hierarchical representations of complex systems. A component model is defined by a set of component and connector types that are instantiated for the system configuration at hand.

The architectural elements of the component model may be annotated with various types of *properties* that provide analytic, behavioral, or semantic information associated with components and connectors. Example properties are the expected throughput and energy consumption of a component, or the expected latency of a connector.

Constraints define restrictions on the system configuration, determining the permitted compositions of elements instantiated from the types. The topological and behavioral constraints determined by the constraints establish an envelope of allowed changes that can be used to ensure the validity of system adaptations.

Analyses are executable specifications that allow examinations of the annotated component model in order to obtain a better understanding of the system. Analysis can be performed on any valid configuration of the component model, including the model of the running system or any reconfiguration that complies with the constraints. An example is a performance analysis of a network based on a queuing network model that allows predicting the latency of traffic in a network for different possible configurations.

Adaptation strategies define the adaptations that can be applied to adapt the system from an undesirable to a desirable state. The adaptations of a strategy can be applied as one atomic set of actions or the strategy can be applied progressively, where the effects of adaptations are observed and continued until a satisfactory situation is obtained.

Adaptation operators are defined using adaptation strategies. Adaptation operators determine a set of actions that the adaptation layer can perform on the elements of a system (components and connectors) to adapt its configuration. Depending on the model element types, the properties, and the constraints, the system may provide operators to add elements, remove elements, replace elements, etc. from a system configuration.

5.3.2 Architecture-based Adaptation of the Web-based Client-Server System

We zoom now in on a concrete realization of architecture-based adaptation with an architectural model. Consider a Web-based client-server system that comprises client and server components that interact remotely through network connections. In particular, Web clients make stateless requests of content from one of a set of Web server groups, as shown in the bottom layer of Figure 5.4, which corresponds to the managed system layer in Figure 5.3. The incoming requests of the clients connected to a server group enter a request queue of the group, where the servers that belong to the group can pick requests and process them.

We consider performance as the primary adaptation concern of the client-server system; in particular we focus on the response time of servers to invocations of clients. Based on an analysis of the system, for instance using a queuing model, the changing number of requests over time, which may be difficult to predict, has been identified as the main factor that affects the response time. Hence, the two main properties of the system that determine the response time are the server load and the available bandwidth. These properties, together with the response time of client invocations, need to be tracked by the probes of the client-server system interface, which corresponds to the managed system interface in Figure 5.3.

The adaptation layer realizes a MAPE-based loop that interacts with the client-server system through the client-server interface. Central to the adaptation layer is the architectural model. The component model represents the client-server system by means of components that represent clients (of type *ClientType*), servers (*ServerType*), server groups (*ServerGroupType*), and links (*LinkType*).

The properties that are relevant for adaptation are the response time of clients (*ClientType.reponseTime*), the server load (*ServerType.load*), and the bandwidth of links (*LinkType.bandwidth*).

The adaptation goal is defined as a constraint for each client (represented as *self*) as follows:

invariant (self.responseTime < maxResponseTime) ! → responseTimeStrategy(self);

This constraint defines an invariant that says that the maximum response time of invocations by a client should not exceed a maximum value.

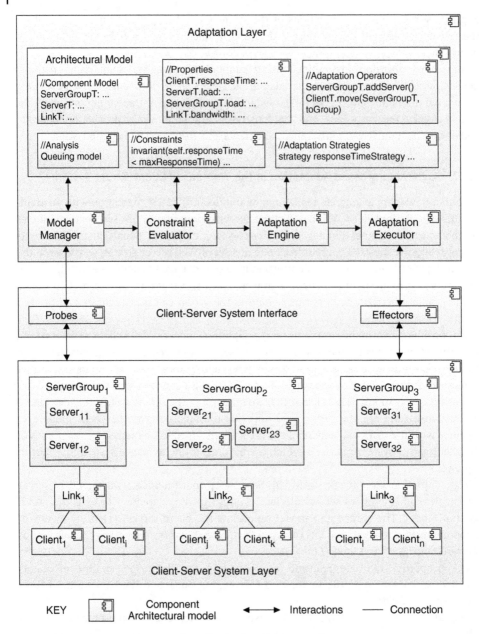

Figure 5.4 Layered architecture of a self-adaptive Web-based client-server system.

If the invariant is violated, the system reacts (indicated by "! →") by applying a response time strategy that is defined as follows:

```
strategy responseTimeStrategy(ClientType C) {
    let G = findConnectedServerGroup(C);
    if (query("load", G) > maxServerLoad) {
```

```
        G.addServer();
        return true;
    }
    let conn = findConnector(C, G);
    if (query("bandwidth", conn) < minBandwidth) {
        let G = findBestServerGroup(C);
        C.move(G);
        return true;
    }
    return false;
}
```

We consider two operators to adapt the system: add a server to a group *(ServerGroupT. addServer())* and move a client from one group to another *(ClientT.move(ServerGroupT, toGroup))*. These operators are supported through the effectors of the client-server system interface.

The model manager keeps track of the response time of clients, the server load, and the bandwidth of links. The constraint evaluator periodically checks whether the measured response time of clients is below a required threshold. If a violation is detected, the constraint evaluator triggers the adaptation engine, which executes the response time strategy. The strategy starts by checking whether the load of the current server group exceeds a predefined threshold *(maxServerLoad)*. If this is not the case, the engine adds a server to the group, decreasing the load and consequently also the response time. However, if the available bandwidth between the client and the current server group drops below a minimum level *(minBandwidth)*, the engine moves the client to another server group. This will result in higher available bandwidth and reduce the response time. Finally, the adaptation executor will apply the required operators to adapt the system, i.e. adding a server to the group if the load of the current server group is too high or moving a client to another server group if the bandwidth between the client and the current server group is too low.

This example shows how architecture-based adaptation is centered around an architectural model, which provides a global perspective on the system. This model allows the definition of system-level properties and integrity constraints that enable a feedback loop to identify violations of the adaptation goals and apply the necessary adaptation actions to the system to resolve the problem. The architecture perspective not only provides an appropriate level of abstraction to reason about the adaptation of the system as a whole, it also fosters reuse as the components of the adaptation layer provide an infrastructure that can be reused across different systems that require adaptation.

5.4 Comprehensive Reference Model for Self-Adaptation

The three-layer model provides a vertical modularization of a self-adaptive system. The model distinguishes the principal concerns of self-adaptation and maps them to different layers. The bottom layer maps to the managed system, which is responsible for the domain concerns. The top two layers map to the managing system. These layers are responsible for the adaptation concerns, which are the focus of self-adaptation. The three-layer model does not focus on horizontal modularization, i.e. modularization within these layers. Both vertical and horizontal modularization are important to support software engineers in (i) defining the key architectural characteristics of complex self-adaptive systems and (ii) enabling the system to reason about change and make effective adaptation decisions. There are two principal ways to approach horizontal modularization: on the one hand, the division into distinct elements that realize the functionality of self-adaptation, corresponding to the MAPE-K model; on the other hand, the division into elements, and their coordination, to realize self-adaptation from a distribution perspective.

In this section, we integrate the different architecture aspects of self-adaptive systems into a comprehensive reference model for self-adaptation. This model is based on FORMS, short for FOrmal Reference Model for Self-adaptation. The integrated model unifies three perspectives. The first perspective, which is based on the principles of computational reflection, focuses on the coarse-grained elements of a self-adaptive system covering the three-layer model. The second perspective, which is based on the MAPE-K model, focuses on the basic functions of self-adaptation and the models these functions share to make adaptation decisions and realize the adaptation goals. The third perspective, which is based on the basic principles of coordination, focuses on self-adaptation in a distributed setting.

5.4.1 Reflection Perspective on Self-Adaptation

Computational reflection is the ability of a software system to modify itself while running. Computational reflection has traditionally been studied at the level of programming languages and usually realized using compiler technologies. The first perspective of the encompassing reference model of self-adaptation is based on the principles of computational reflection. However, we apply these principles at the architectural level to modularize self-adaptive software systems into a set of interacting basic elements.

Aligned with the basic principle of computational reflection, we distinguish between models and computations as the constituent elements of a software system (or a sub-system). Intuitively, a model comprises representations, which describe something of interest in the physical and/or cyber world, while a computation is an activity in a software system that manages its own state. The computation of a software system can reason about its models, act upon them, and interact with external elements.

Figure 5.5 shows the reflection perspective of the reference model in the Unified Modeling Language notation (UML).

A *Self-Adaptive System* is situated in an *Environment*. The environment consists of *Attributes* and *Processes*. An attribute is a perceivable characteristic of the environment. A process is an activity that can change the environment attributes. For instance, attributes for DeltaIoT are the motes, the gateway, and the network links, while the noise in the environment that may cause interference along links is a process. In general, the environment may correspond to both physical and logical entities. Therefore, the environment of a computing system may itself be another computing system. For example, the environment of DeltaIoT includes both the gateway and the software of the management interface that computes the statistical data about the quality of service of the network.

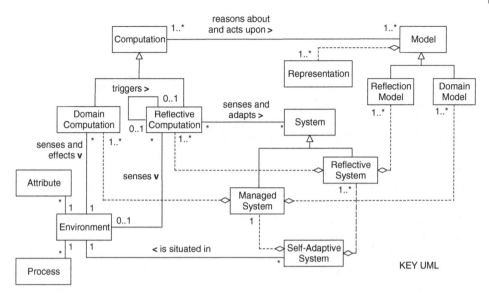

Figure 5.5 Reflection perspective of the encompassing reference model for self-adaptation.

The distinction between the environment and the self-adaptive system is made based on the extent of control. Everything relevant to the self-adaptive system that is not under control of the system is part of the environment. For instance, in DeltaIoT, the self-adaptive system may interface with intelligent sensors that track the SNR along the links of the network, but since the system cannot manage or adapt their functionality, these sensors are part of the environment.

A self-adaptive system consists of a *Managed System* and one or more *Reflective Systems*. The managed system realizes the goals for the users of the self-adaptive system. For DeltaIoT, the managed system provides the building security services to the users. The managed system comprises a set of *Domain Models* and a set of *Domain Computations*. A domain model represents something of interest in the domain of the application. Domain models in DeltaIoT may incorporate a variety of information, such as the network topology, the types of sensors deployed on the motes, etc. Domain computations sense the environment, reason about and act upon domain models, and affect the environment. Domain computations in DeltaIoT sense parameters in the environment, store the sensor data in a queue, transmit packets over the network according to the time schedule, etc.

A reflective system is a part of the software system that manages another system, which can be either a managed system or another reflective system. Hence, reflective systems may manage other reflective systems forming a hierarchy. The combined set of reflective systems of a self-adaptive system makes up the managing system as defined in the conceptual model of a self-adaptive system. For a self-adaptive system with a single reflective system, this reflective system will manage the managed system. For a self-adaptive system with a hierarchy of two reflective systems, the one at the bottom will manage the managed system; the one at the top will manage the reflective system at the bottom. For instance, one reflective system may manage the adaptation of network settings to realize the required packet loss and latency goals. Another reflective system on top of this may learn over time the impact of adaptation decisions in particular situations and adapt the decision-making process of the first reflective system to improve the adaptation results.

A reflective system comprises a set of *Reflection Models* and a set of *Reflective Computations*. A reflection model represents entities of the layers beneath, such as elements of a system and environment attributes. Analogous to domain computations, reflective computations maintain reflection models, reason about the content of the models, and act upon them. For instance, a reflection model for DeltaIoT may be an architectural model that represents the actual configuration of the network, which is used at runtime by a reflective computation to reason about adaptation and adapt the network settings or rewire links between the motes.

It is important to note that a reflective computation may also monitor the environment to determine when adaptations are necessary. For instance, a reflective computation in DeltaIoT may track the SNR levels of the links to determine how the network needs to be adapted. However, unlike the domain computation, a reflective computation is not allowed to affect the environment directly. This essential principle of computational reflection ensures a clear separation of concerns (also called a disciplined split): domain computations are concerned with the environment; reflective computations are concerned with either the managed system or another reflective system.

The reflection perspective of the encompassing reference model for self-adaptation delineates the boundaries between various key elements of self-adaptive systems. In particular, the perspective clearly distinguishes between elements that constitute the environment, the managed system, and the reflective system(s), together with their responsibilities and constraints.

5.4.2 MAPE-K Perspective on Self-Adaptation

The reflection perspective tells us that a reflective system is a self-contained entity that adapts another system, being either the managed system or a lower level reflective system, using reflective computations that reason about and act upon one or more reflection models. The MAPE-K perspective refines the abstract notions of reflective computation and reflection model into concrete basic elements, i.e. the essential computations and models.

Figure 5.6 shows the MAPE-K perspective of the encompassing reference model. The elements marked in gray shaded boxes denote the integration with the reflection perspective.

Aligned with the primary functions of self-adaptation, the computations of a reflective system are the feedback loop computations found in self-adaptive systems: monitor, analyze, plan, and execute. These computations employ four types of reflection models: the subsystem model, environment model, adaptation goals model, and MAPE working model. The combined reflection models correspond to the Knowledge that is used by the MAPE functions. The different types of computations and models are illustrated for DeltaIoT in Section 4.4.1.

It is important to note that the monitor computation of any reflective system has the possibility to sense the environment, the managed system, and if applicable the underlying reflective systems. For any of these, the reflective system may provide corresponding reflection models. However, the execute computation of any reflective system is only allowed to adapt the system in the layer beneath. On the one hand, this offers the reflective

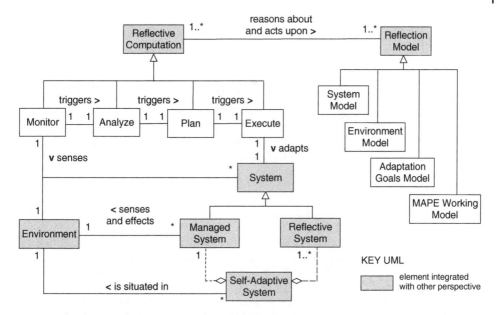

Figure 5.6 MAPE-K perspective of the encompassing reference model for self-adaptation. The elements marked in gray shaded boxes denote the integration with the reflection perspective.

computations the flexibility to take into account the context in which the underlying system (managed and reflective) operates in order to make well-informed adaptation decisions. On the other hand, this constraints the scope of allowed adaptation actions ensuring a clear separation of concerns. That is, each reflective system is only concerned with the layer beneath and realizes the adaptation goals for the system in that layer.

5.4.3 Distribution Perspective on Self-Adaptation

The distribution perspective focuses on the modularization of a self-adaptive system within layers across computational elements. This modularization is particularly important for engineering distributed self-adaptive systems, which make up the majority of real-world systems. The distribution perspective explicitly separates coordination from computation.

In a distributed system, the software is deployed on various computational elements (nodes) that are connected via a network. Such a setting requires dedicated coordination mechanisms that enable the software running at the different computational elements to work together to realize the system goals. In general, a coordination mechanism allows a distributed system to resolve problems that arise from dependencies in the system, such as dependencies between tasks and resources. A wide variety of coordination mechanisms for computing systems exist. Examples are master-slave and market mechanisms. The choice of a particular mechanism depends on the coordination problem at hand and the characteristics of system and its environment.

Equipping a distributed system with self-adaptation capabilities requires proper support for coordination of the reflective computations that deal with adaptation. Figure 5.7 shows the distribution perspective of the reference model integrated with the other perspectives.

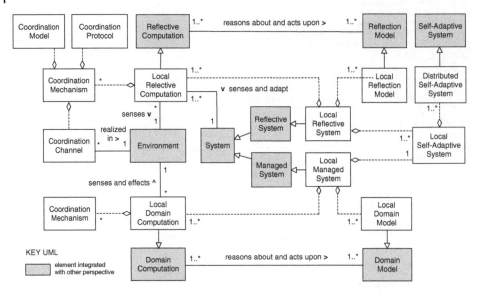

Figure 5.7 Distribution perspective of the encompassing reference model for self-adaptation. The elements marked in gray shaded boxes denote the integration with the other perspectives.

A *Distributed Self-Adaptive System* consists of multiple *Local Self-Adaptive Systems*. Typically, each local self-adaptive system is deployed on a node of a network. Consider an extended setup of DeltaIoT as shown in Figure 5.8. This configuration comprises two subnets, each with one gateway. The software deployed on each gateway constitutes a local self-adaptive system.

A local self-adaptive system comprises a *Local Managed System* and one or more *Local Reflective Systems*. A local managed system provides the domain functionality of one computational element of the distributed system. A local managed system comprises *Local Domain Computations* and *Local Domain Models*. A local domain computation extends a domain computation with a *Coordination Mechanism*. The coordination mechanism allows the local domain computations to coordinate with each other when needed. In the DeltaIoT scenario, the local managed systems deployed at the gateways collect sensor data of the motes in their respective subnets. The motes at the border of the subnets (marked in gray shading in the figure) may send sensor data to both gateways.

A local reflective system manages another system, which can be either the local managed system or another local reflective system. A local reflective system comprises *Local Reflection Computations* and *Local Reflection Models*. In DeltaIoT, local reflective systems deployed at the gateways are responsible for deciding how to divide the fraction of packets that are sent by these border motes to each subnet. Depending on the traffic load generated in the subnets and other conditions (e.g. the local levels of interference), these fractions will affect the quality properties of the global netwerk. More precisely, the local reflective systems need to determine the values of α_{moteID} and β_{moteID} such that:

$$\alpha_{moteID} + \beta_{moteID} = 1$$

with α_{moteID} the fraction of packets sent by border mote with identifier *moteID* to the subnet of *gateway 1* and β_{moteID} the fraction of packets sent by the mote to the subnet of *gateway 16*.

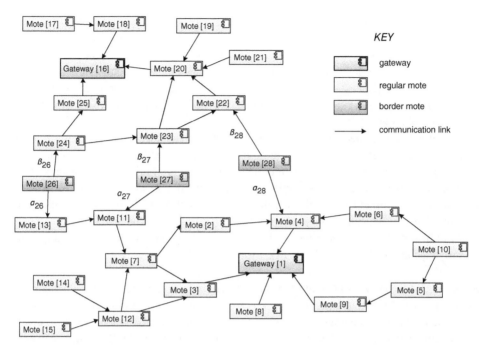

Figure 5.8 DeltaIoT scenario with distributed self-adaptation. The network consists of two gateways, each collecting sensor data of a subset of motes as indicated by the communication links.

A local reflective computation extends a reflective computation with a *Coordination Mechanism*, which allows the computation to coordinate with other local reflective computations in the same layer. In the DeltaIoT scenario, the local reflective computations at the gateways have to coordinate to determine the fraction of packets the border motes have to sent to each subnet. Such a coordination can be organized using a master-slave mechanism with one gateway being responsible for determining the distribution of packets periodically. Alternatively, a peer-to-peer mechanism may be used where both gateways can initiate the coordination to revise the distribution of packets when needed.

A coordination mechanism is a composite consisting of a *Coordination Model*, a *Coordination Protocol*, and a *Coordination Channel*, which is a commonly accepted structure for coordination mechanisms.

A coordination model contains the information that is used by a local reflective computation to coordinate with reflective computations of other local managed systems. A coordination model may represent information about the partners involved in the coordination with their roles, status information about the ongoing interactions, etc. In the DeltaIoT scenario, the local reflective computations at the gateways that determine the fractions of packets sent by the border motes to the subnets may use a coordination model that registers their roles, the current fractions that the border motes use to send packets, the average values of packet loss over a recent time window, and the latency and energy consumption of the subnets, among other data.

The coordination protocol represents the rules that govern the coordination among the participating computations. Examples are voting in peer-to-peer coordination and an auction in market-based coordination. When master-slave coordination is used in DeltaIoT to decide about the fractions of packets sent to the subnets by the border motes, the gateways may use a simple call-return protocol, where the master gateway requests information from the slave gateway, computes new values for the fractions if needed, and if so, instructs the slave gateway to apply the new schedule. For instance, the fraction of packets sent to one of the subnets may be increased if the other subnet temporarily suffers from high packet loss.

A coordination channel is a semantic connector that acts as the means of communication between the parties involved in a coordination. A coordination channel can be an abstraction for direct interactions (regular communication channels for message exchange) as well as indirect interactions (e.g. shared tuple spaces). A coordination channel is eventually realized in the environment in which the self-adaptive system is deployed. For example, the coordination channel of a master-slave coordination mechanism that relies on a call-return protocol in DeltaIoT may use direct exchange of messages, which can be realized via the underlying communication and network infrastructure between the gateways.

A Note on Distribution versus Decentralization   With distribution we refer to the deployment of the software of a self-adaptive system to hardware. Hence, distribution of a self-adaptive system refers to the deployment of the software of both the managed system and the managing system on computational elements. The opposite of a distributed self-adaptive software system is a system for which the software is deployed on a single computational element (node). The deployment of the managed system and the managing system do not have to coincide: the software of a managed system may be deployed on a set of computational elements, while the software of the managing system may be deployed on one dedicated computational element. In this case, the managed system is distributed and the managing system is not.

With decentralization we refer to how adaptation decisions in a self-adaptive system are made and coordinated among different local reflective systems, independently of how those systems are physically distributed. A local reflective computation is able to make local adaptation decisions and coordinate with other local reflective computations to realize the overall adaptation goals. The extent to which local reflective systems are able to make independent decisions determines the degree of decentralization of decision making of a distributed self-adaptive system.

At a more fine grained level, decentralization can also be considered in terms of the four basic functions of self-adaptation: monitoring, analyzing, planning, and execution. These functions can have different degrees of decentralization. For instance, analysis can be centralized, meaning that a single analysis computation is responsible for performing the analysis for the adaptation of the whole system, while planning may be completely decentralized and performed locally by local planning computations.

The degree of decentralization of adaptation decision making is independent of how the software of the self-adaptive system is deployed. However, in practice, when the software of the managed system is deployed on a single computational element, the software of the managing system will often also be deployed on that element, and the adaptation functions are typically centralized. Similarly, managing systems with fully decentralized adaptation functions typically go hand in hand with the distribution of the software of the managed and managing systems. Between these two extremes, there is a range of different ways to organize the deployment and decision making of a self-adaptive system.

5.5 Summary

Architecture-based adaptation is concerned with the basic principles of systematically engineering self-adaptive systems. Architecture-based adaptation has two facets. On the one hand, it offers architectural abstractions to define self-adaptive systems; on the other, it offers architectural models that enable the system to reason about adaptation and make adaptation decisions.

From a system definition point of view, an architectural perspective to self-adaptation offers a vehicle to clearly separate the domain concerns from the adaptation concerns, which adds to the understandability, maintenance, and reuse of self-adaptive systems. Architecture offers an integrated approach that spans a wide range of engineering activities, and it enables leveraging consolidated engineering efforts, such as notations and analysis methods.

From the point of view of reasoning about adaptation, an architectural perspective offers an appropriate level of abstraction to manage change and deal with system-wide concerns and constraints. Furthermore, architecture facilitates scalability through composition.

The three-layer model of self-adaptive systems, which is inspired by a classic architecture of robotic systems, provides a holistic view on self-adaptation by distinguishing the main high-level activities involved in self-adaptation and the allocation of concerns to these activities.

The bottom layer of the three-layer model is component control, which consists of the constituent components that realize the goals for the users of the system. The middle layer, change management, can enact plans to adapt the underlying system in order to deal with problems reported by the system. The top layer is goal management, which produces plans based on high-level goals. Change management can request such plans if it faces situations it cannot handle. Stakeholders can add plans when new goals are added, or the plans may be synthesized automatically – but this is a complex task.

Component control, the bottom layer of the three-layer model, maps to the managed system in the conceptual model of a self-adaptive system. Both change management and goal management map to the managing system of the conceptual model. The three-layer model makes an explicit distinction between the functionality of the managing system that is responsible for realizing the adaptation goals in the current situation – i.e. change management – and the functionality that is responsible for selecting adaptation goals over time when the conditions of the system change – i.e. goal management.

Architecture-based adaptation exploits an architectural model that represents the system as a composition of components, their interactions, and their properties of interest. An architectural model exposes important system level properties and integrity constraints, facilitating decision making for self-adaptation using a global perspective on the system.

Rainbow offers an established framework for the realization of architecture-based adaptation using an architectural model. The framework realizes the basic functions of a managing system that share an architectural model. This architectural model comprises

Table 5.1 Key insights of Wave II: Architecture-based Adaptation.

• Architecture-based adaptation has two perspectives that concern respectively: the definition of self-adaptive systems and the decision-making of adaptation during operation.

• Architecture provides a foundation to manage the complexity of defining a self-adaptive system.

• An architectural model offers an appropriate representation of the system enabling a feedback loop to reason about change and make adaptation decisions.

• Two fundamental architectural concerns of self-adaptation are change management (i.e. managing adaptation using plans) and goal management (i.e. generating plans based on high-level goals).

• An architectural model exposes important system level properties and integrity constraints, facilitating decision making for self-adaptation using a global perspective on the system.

• Three primary and interrelated perspectives on self-adaptive systems are: reflective computation, MAPE-K, and distributed coordination.

---

a component model of the managed system, properties associated with architectural elements, constraints that determine the envelope of allowed changes, specifications to analyze valid system configurations, and adaptation strategies and adaptation operators that can be applied to adapt the system and solve the adaptation problem.

The integrated reference model for self-adaptation unifies three perspectives. The reflection perspective modularizes the main basic elements of a self-adaptive system aligned with the principles of computation reflection. The MAPE-K perspective identifies the basic computations of a self-adaptive system and the basic models they reason about and act upon. The distribution perspective deals with the modularization of a self-adaptive system within layers and across the nodes of a distributed self-adaptive system. Central to this modularization is the coordination among computations that realize self-adaptation and the way adaptation decisions are made (from centralized to decentralized).

Table 5.1 summarizes the key insights of Wave II.

5.6 Exercises

5.1 Three-layer model applied to EUREMA: level H

EUREMA is a model-driven engineering approach to realizing self-adaptive software. Study EUREMA and apply the three-layer model for self-adaptation to this approach. Discuss the elements for each layer of the instantiated model and their responsibilities, and explain the flow of activities across the layers. Assess the EUREMA approach in light of the three-layer model for self-adaptation. For the EUREMA paper, see [195]; for the paper with the three-layer model, see [121].

5.2 Reflection and MAPE-K perspectives for QoSMOS: level H

QoSMOS is a formally-founded approach that can be used to develop service-based systems that achieve their QoS requirements through dynamically adapting to changes in the system state, environment, and workload. Study QoSMOS and instantiate the reflection and MAPE-K perspectives of the comprehensive reference model of self-adaptive systems for the approach. Discuss the responsibilities of each element and explain the collaborations between the elements. Assess the QoSMOS approach from the viewpoint of the comprehensive reference model. For the QoSMOS paper, see [39].

5.3 Feedback loop functions of the MUSIC platform: level D

MUSIC is a component-based planning platform that optimizes the overall utility of applications that are subject to changing operating conditions. The platform allows interchangeable components and services to be plugged in automatically when the execution context changes. Study the MUSIC platform and map its architecture to the basic architecture of architecture-based adaptation with an architectural model. Identify the feedback loop functions together with the architectural models they share. Discuss the responsibilities of these elements, and explain the flow of activities that occur during adaptation. Assess the MUSIC approach in relation to architecture-based adaptation with an architectural model. For the MUSIC paper, see [167].

5.4 Utility functions in SWIM: level D

SWIM is an exemplar for self-adaptation in the domain of Web applications. Servers in SWIM are simulated, which makes experimentation with adaptation managers easy. Download the SWIM exemplar and install it. Configure the adaptation manager as a simulation module and use the standard configuration file that comes with the artifact. Run the artifact in simulation mode and analyze the results. Adjust the utility function used by the adaptation manager and study the effects on the adaptation results. Assess the results of the experiments. For the SWIM artifact, see [143]; for the accompanying paper, see [142].

5.5 Decentralized adaptation in TAS: level W

Setting. Consider a large-scale setting of the health assistance system that we introduced in Chapter 1 where two providers offer health services to patients, each one responsible for a geographic area. Each health service system is an independent entity that uses concrete services. Some of the services are obtained from public service providers; these services can be used by both health service systems. Other services are obtained from providers that offer private services to one particular health service system. To ensure a good overall service to their clients, the health service systems work together. The challenge for self-adaptation is to enable both systems to provide the required reliability of services to clients, while minimizing the cost at all times. There may be peak demands in service requests for one or both systems. Furthermore, the systems should take into account that services may disappear at any time and new services may emerge. To deal with these challenges, the entities need to make decisions about which public and private services to select. Such selections may be revised periodically or when particular events occur at runtime. A key aspect for the two providers of the health services will be to agree on the public services they use. The providers may also exchange load (service requests) when one of them faces a peak in demand.

Task. Design a managing system that enables the two health service systems to adapt in a collaborative manner. Identify what data the two entities need to exchange and what protocol they can use to coordinate the adaptations. Instantiate the distribution perspective of the comprehensive reference model for self-adaptation for your solution. Extend the TAS artifact to implement and evaluate your solution.

Additional material. Information about the TAS artifact is available at the TAS website [201]. The patterns described in [209] offer inspiration to solve the adaptation problem.

5.6 Self-adaptive agent system for the Packet-World: level W

Setting. The Packet-World is a test bed to experiment with multi-agent systems. The Packet-World consists of a number of differently colored packets that are scattered over a rectangular grid. Agents that live in this virtual world have to collect these packets and bring them to the correspondingly colored destinations. Figure 5.9 shows an example of a Packet-World of size 10 × 10 with 8 agents. Shaded rectangles symbolize packets that can be manipulated by the agents and circles symbolize destinations. Agents can make a step to a free neighboring cell. If an agent is not carrying any packet, it can pick up a packet from one of its neighboring cells. An agent can put down a packet it carries at the destination point of that particular packet, but also at one of the free neighboring cells. If there is no sensible action for an agent to perform, it may wait for a while and do nothing. Agents can also exchange messages; in particular, agents can request information about packets or destinations from each other. The sensing-range of the world expresses how far, i.e. across how many squares, an agent can perceive its neighborhood. Figure 5.9 illustrates the limited view of agent 8: in this example the sensing-range is 2. Similarly, the communication-range determines the scope within which agents can communicate with one another. The goal of the agents is to perform their job efficiently – i.e. with a minimum number of steps, packet manipulations, and message exchanges.

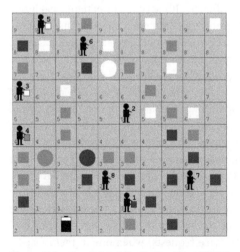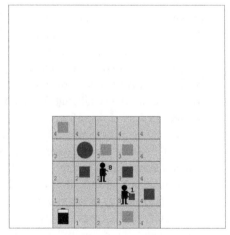

Figure 5.9 Example of the Packet-World [204].

Task. Download the Packet-World. Get familiar with the test bed by experimenting with a number of predefined setups. Now instantiate a setup of the Packet-World as shown in Figure 5.9, but without the battery. Give the agents a sensing-range of 2 and a communication-range of 3. Design an agent architecture to solve the problem. Implement your design, instantiate it for each agent, and test the solution. Now equip each agent with a battery, and give the agents an initial level of energy. Also add a battery charger to the environment that emits a field that agents can sense, similar to that shown in Figure 5.9 (using the numbers in the bottom-left corner of the fields). Extend your basic agent architecture now with a managing system that enables each agent to check its battery level and activate behavior to recharge the battery at the charger station when needed. Implement your extended design, instantiate it, and test your solution. Assess your solution.

Additional material. The Packet-World can be downloaded via [64]. The test bed offers different predefined agent implementations that can be used as inspiration. For more information about the Packet-World, see [204].

5.7 Bibliographic Notes

M. Gorlik and R. Razouk introduced the architectural style Weaves [89], which supports dynamic rearrangement of workflows. J. Magee et al. introduced the architectural description language Darwin, which can be used to define hierarchical compositions of interconnected components that may evolve dynamically during execution [132].

D. Garlan et al. emphasized that external mechanisms offer a more effective solution to self-adaptation compared to internal mechanisms [81]. P. Maes pinpointed the notion of computational reflection, referring to a system that incorporates structures representing aspects of itself [129]. These structures provide a causally connected self-representation of the system, which makes it possible for the system to answer questions about itself and support actions on itself. G. Kiczales et al. combined object-orientation and reflective techniques into a meta-object protocol [115]. A meta-object protocol implements an object system using its own constructs as building blocks, adding flexibility to a programming language. The principles of meta-object protocols laid the foundation for aspect-oriented programming, which provides basic programming languages with constructs to modularize design decisions that would otherwise be scattered throughout the code [116].

J. Kramer and J. Magee argued for an architectural perspective on self-adaptation and introduced the three-layer model for self-adaptive systems [121]. This model leveraged the classic three-layer model of E. Gat that is widely used in robotic systems [82]. N. Villegas et al. presented the DYNAMICO model for engineering adaptive software, which puts the emphasis on mechanisms required to deal with changing adaptation goals [192].

P. Oreizy et al. highlighted the relevance of an architectural model to reason about the underlying executing system and adapt it during operation [151]. D. Garlan et al. introduced the basic architecture of a feedback loop that reasons over an architectural model. The authors implemented this architecture in Rainbow, a classic framework for realizing architecture-based adaptation [81]. Rainbow was applied to the Web-based client-server system described in this chapter.

The comprehensive reference model for self-adaptation is based on FORMS [206]. The reflection perspective is inspired by the principles of computational reflection [129]. The MAPE-K perspective is grounded in the seminal work of J. Kephart and D. Chess [112]. T. Malone et al. performed seminal work on the principles of coordination mechanisms [138]. Well-known coordination mechanisms are tuple spaces [83] and protocols [110]. The difference between distribution and decentralization in the context of self-adaptation is discussed in [209].

There are many online resources available about modeling with UML. M. Seidl wrote an instructive book about UML modeling [173].

6

Wave III: Runtime Models

The third wave puts the concrete realization of self-adaptation in focus. The second wave clarified the architectural principles that underlie the definition of self-adaptive systems and the means for reasoning about adaptation. While these principles provide an integrated high-level perspective on how to engineer self-adaptive systems, the concrete realization remains complex. In particular, the second wave emphasized the need for an appropriate representation of the system in the form of an architectural model that enables the system to reason about change and make adaptation decisions. While this is pivotal, a self-adaptive system in general requires various types of models that represent different aspects of self-adaptive systems to support the decision-making of adaptation.

Leveraging the insights from the second wave, in the third wave the focus shifts to the models used by a self-adaptive system during operation, in particular the different types of runtime models and their dimensions, the role runtime models play, and the ways in which the models can be used in the realization and decision-making of self-adaptation. Understanding these different facets of runtime models is essential for engineers to manage the complexity of the concrete design and implementation of self-adaptive systems.

Runtime models can be considered as an extension of model-driven engineering into the runtime. Model-driven engineering is a software development approach that emerged in the 1980s. The approach is centered on the reuse of standardized models for the design of domain models in close interaction with the stakeholders. The basic idea is to apply incremental refinement of the models, which will eventually enable the implementation of the software system. Model-driven architecture is a well-known incarnation of model-driven engineering, which leverages established notations, standards, domain expertise, and tools to specify models for the development of software systems. Whereas model-driven engineering is centered on the offline activities of engineers that realize a software system through model refinements using tools, runtime models extend this to the runtime and are centered on the activities of a system that uses up-to-date models to reason about itself, its context, and its goals, and to adapt autonomously in order to deal with the changing conditions it encounters during operation.

Runtime models provide an answer to the need to manage the complexity of concrete designs of self-adaptive systems, which was the driver for the third wave. The third wave is concerned with the software models specified during the development of a self-adaptive system, possibly incomplete, and the role these models play during the operation of the

An Introduction to Self-Adaptive Systems: A Contemporary Software Engineering Perspective,
First Edition. Danny Weyns.
© 2021 John Wiley & Sons Ltd. Published 2021 by John Wiley & Sons Ltd.

system. Incomplete models are updated at runtime based on observations of the executing system and its environment.

In this chapter, we provide a definition of a runtime model in self-adaptive systems. We elaborate on the motivations for runtime models, and we present different dimensions of runtime models. Then, we describe three principal strategies that can be applied to using runtime models in self-adaptive systems.

LEARNING OUTCOMES

- To explain what a runtime model is and what weak causal connection means.
- To motivate the use of runtime models in self-adaptive systems, illustrated with examples.
- To explain and illustrate different dimensions of runtime models.
- To summarize and compare the principal strategies for using runtime models in self-adaptation.
- To explain each principal strategy in detail and illustrate it with an example.
- To identify the principal strategy applied in a concrete self-adaptive system.

6.1   What is a Runtime Model?

In model-driven engineering, a model is an abstraction of a system that is built for specific purposes. A runtime model is aligned with this view, but extends it in two principal ways. First, a runtime model can be an abstraction of a system *or any aspect related to the system*. Second, a runtime model that is specified during the development of the system becomes *a first-class citizen at runtime*. Applied to a self-adaptive system, we define a runtime model as follows:

> A runtime model of a self-adaptive system is a first-class runtime abstraction of that system or any aspect related to that system that is used for the purpose of realizing self-adaptation.

In Chapter 5, we distinguished four types of models that are maintained in the knowledge of a managing system, respectively representing the managed system, the environment, the adaptation goals, and working data of the managing system. These representations form the main types of runtime models that are used in self-adaptive systems. In this chapter, we elaborate on how these model types can be instantiated and can be used in different ways as runtime models by the feedback loop components (which realize the MAPE functions). Besides the representation of knowledge shared by the feedback loop components, runtime models can also be used to realize the MAPE functions; we elaborate on this further in this chapter.

6.2   Causality and Weak Causality

Since a runtime model represents the system or aspects of it that are relevant to self-adaptation, consistency between the model and the represented system or aspects of it is an important property. In particular, since runtime models are runtime entities that are part of, or related to, a system that is subject to continuous change, it is essential that the state of the models is kept up-to-date. More specifically, if a runtime model represents a system or aspects of it that change dynamically, that representation should mirror the changes at any time. A managing system that uses runtime models that do not reflect the artifacts they represent may take inferior adaptation decisions. Worst-case, such decisions may lead to system failures or even catastrophes, in the case of a system with strict goals.

As we have seen in Chapter 5, in computational reflection, a system maintains a representation of itself. An essential property of such a self-representation is a causal connection between the representation and the underlying system. The causal connection ensures that if the system changes, the representation of the system also changes, but also if the representation changes, the system changes accordingly.

Causal connection as defined in computational reflection is a strong property. Causal connection requires a bi-directional synchronization between the representation, i.e. a model, and the system or aspects of the system that the model represents at any time. Causal connection might be realized for runtime models of a self-adaptive system that represent (parts of) the managed system, if that system is fully controllable. However, there are different reasons why causality as defined in computational reflection cannot always be achieved for runtime models of self-adaptive systems. For instance, a runtime model may represent an aspect relevant to the system that is not fully controllable. A simple example is a property of the environment. A runtime model may also represent an aspect of the system or its environment that requires sensing and processing of data, which may cause a delay in the updates of models. Similarly, it may not be possible to instantaneously adapt a system when the corresponding model changes, because adaptation of the system may require time. Furthermore, runtime models may not represent the ground truth when dealing with uncertainty. In such cases, it may be hard or even infeasible to keep models and aspects of the system in perfect synchrony.

Hence, we introduce a weak version of causal connection for runtime models of self-adaptive systems, which is loosely defined as follows:

> A weak causal connection links the state of a runtime model and the state of a self-adaptive system or relevant aspects of it, allowing an acceptable discrepancy between the two; the discrepancy can be temporal, quantitative, or take any other form of variation.

By acceptable discrepancy, we mean that the discrepancy should be minimized and in any case not invalidate the decision-making process of the managing system.

6.3 Motivations for Runtime Models

The motivation for the use of runtime models is multi-fold. We list the main arguments that have been put forward in favor for the use of runtime models.

- *Representing dynamic change*: Runtime models provide an abstraction of the system or aspects related to it. Abstraction enables dynamic change to be described at system-wide level; it is essential to focus on the runtime facets that are relevant to adaptation and omit details. As such, runtime models extend the use of an architecture model to reason about self-adaptation, from Wave II.

- *Separation of concerns*: Distinct models can capture different aspects of the system. This separation of concerns paves the way for various use cases for runtime models: for instance to track and understand the runtime behavior of the system or its environment, to maintain historical information about the system, to represent the adaptation space during operation, and to predict the effects on the system goals of applying different adaptation options.
- *Runtime reasoning*: Monitoring the system or aspects related to it and using the collected data to keep the runtime models up-to-date enables the managing system to reason about system-wide properties at runtime. Runtime models enable different types of reasoning, from checking constraints over a runtime model and using simulations to perform "what if" analysis to formal analysis to provide guarantees about the expected outcomes of adaptation actions.
- *Leveraging humans in the loop*: Runtime models are abstractions of the running system that can be designed such that human stakeholders can understand them. Human readability of models requires proper representations, for instance models can be specified in domain-specific languages, or they can be generated automatically from models used by the system that may be annotated with semantic information. As such, runtime models can leverage the involvement of humans in different functions of self-adaptation: e.g. to add data to runtime models, to evaluate models (or support the system with evaluating models), to support the decision-making based on the models, and to help guide the execution of an adaptation plan of a large-scale system.
- *Facilitating on-the-fly evolution*: Runtime models are work products of system design that are kept alive at runtime and used by the system. Runtime models can be queried during operation to gain information about the system and detect situations that require an evolution of the system – e.g. to detect a bug or an inefficiency. When a self-adaptive system goes through an evolution step, the runtime models may evolve as well. When these evolved models are self-contained entities, they can then be deployed through live updates.

This list highlights a number of important arguments that underpin the motivation for the use of runtime models in self-adaptive systems. The list is not complete – researchers and engineers have brought forward other arguments in favor of runtime models. Examples are support for semantic integration of heterogeneous elements of a system, and automatic synthesis of models at runtime. An example use case that combines these two is on-the-fly generation of a connector that allows two components with similar but not identical interfaces to interact with one another. Such a connector may be modeled in the form of two communicating state machines.

6.4 Dimensions of Runtime Models

Runtime models can be classified according to different dimensions. Each dimension describes a particular characteristic of the representation of a self-adaptive system or aspects related to the system. Dimensions provide engineers with a common set of options to specify the runtime models for a self-adaptive system under consideration and select suitable representations. It is important to note that the dimensions are to some degree orthogonal – that is, when specifying a runtime model, the engineer can select an option for each of the dimensions. However, in general, not all dimensions apply to all runtime models. In that case, the engineer can ignore the dimensions that do not apply.

We present four key dimensions of runtime models. Each of these dimensions is characterized by a domain with two options. Note that these two options per dimension represent the extremes of the domain. In practice an engineer may apply an option in between these extremes. Furthermore, a practical self-adaptive system typically employs multiple runtime models that combine different dimensions to capture different aspects of the system or aspects related to the system.

6.4.1 Structural versus Behavioral

In general, models typically focus either on the structure or on the behavior of a system or its related aspects. This distinction also applies to runtime models.

Structural models show the actual structure of the system or related aspects. The focus is on the composition of elements and their relationships, which can be at the level of types or of instances. Structural models can be defined at different levels of abstraction, for instance a course-grained model of the deployment of subsystems to nodes of a network or a more fine-grained component and connector model of the configuration that is deployed on a node. The elements in a structural model may represent any element or aspect that is meaningful to self-adaptation.

Behavioral models show the dynamic behavior of a system or parts of it, typically in relation to changes over time. Behavioral models can describe a variety of elements or aspects that are useful to self-adaptation. This includes aspects related to activities, such as series of events generated in the environment, control flows of computations, and flow of data. Behavioral models can also describe aspects related to state changes, such as transitions of the state of a system based on conditions, and protocols used by interacting elements. Other types of behavioral models describe the aspects of the interaction of elements, for example a protocol that describes the order in which messages are exchanged.

Example   Figure 6.1 illustrates a structural model and a behavioral model of an individual mote in DeltaIoT.

The structural model shows the different queues and the manager of a mote, and how they are connected with one another and with external elements. The behavioral model shows the state transitions of the node manager, which represent the flow of activities. These models can be used by the MAPE components to identify the state of a node and query the content of the queues. Such data can, for instance, be used to determine the latency of the network generated by specific motes.

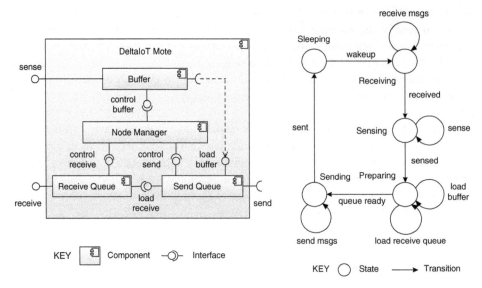

Figure 6.1 Illustration of a structural model (left) versus a behavioral model (right) for DeltaIoT.

6.4.2 Declarative versus Procedural

Another key choice engineers of runtime models need to make is to decide whether models are declarative or procedural. These two types of models complement one another, and both are usually required in the realization of a self-adaptive system.

A declarative model represents knowledge about *what* needs to be done, or the knowledge simply states *that* something is the case. For instance, an adaptation goal specifies the purpose of self-adaptation, i.e. what the managing system should realize. A procedural model, on the other hand, represents knowledge about *how* something is done, or can be or needs to be done. For instance, a plan describes how a managed system needs to be adapted.

Example Listing 6.1 shows small excerpts of a declarative model and a procedural model in DeltaIoT.

The top part defines the adaptation goals for DeltaIoT. The specification declares that the maximum loss of packets sent in the network should not exceed 10 %. The bottom part defines a simple procedure to test whether the packet loss goal is satisfied or not for a given configuration *(gConf)*.

```
     Listing 6.1: Excerpts of declarative model and a procedural model in DeltaIoT.
// Declaration of the adaptation goals
int MAX_PACKET_LOSS = 10; //max packet loss 10%
...
// Procedures to test the adaptation goals for a given configuration
bool satisfactionGoalPacketLoss(Configuration gConf,
   int MAX_PACKET_LOSS) {
      return gConf.qualities.packetLoss < MAX_PACKET_LOSS;
}
...
```

6.4.3 Functional versus Qualitative

A runtime model can represent functionality of a self-adaptive system or it can represent qualitative aspects of the system or a related aspect of it. A self-adaptive system needs both functional and quality runtime models, but often the emphasis is on quality models.

6.4.3.1 Functional Models

A functional model is a structured representation of functionality within the modeled system or its domain. Functionality can be described in the form of elements that perform functions, together with their interrelationships. Functional models can represent the internals of a system or an element of them, input or output of elements, or flows between elements. Functional models can be used to perform a functional analysis of a system or a process. Such an analysis can, for example, be used to check whether a reconfiguration of a self-adaptive system is valid, meaning that the system after adaptation will provide all the required services.

Example Figure 6.2 shows a simple functional model of the effector of the DeltaIoT network.

The left part of the figure shows the effector component with the functions it offers via two provided interfaces. The *setNetworkSettings* interface can be used by the managing system to adapt the managed system with new mote settings. The *setDefaultSettings* interface can be used to adapt the mote settings to default values, for instance to bring the network to a fail-safe operation mode. The realization of these two functionalities relies on the *setMote* interface, which needs to be provided by the managed system. The state diagram on the right shows the flow of activity when the *setNetworkSettings* function is called. The effector adapts the settings of the motes one by one. The effector is notified by the managed system when the adaptation of a mote completes. The effector notifies the managing systems when the adaptation of the whole network is completed.

6.4.3.2 Quality Models

Runtime quality models are essential artifacts for a self-adaptive system as qualities lie at the heart of self-adaptation. In particular, a self-adaptive system requires runtime quality

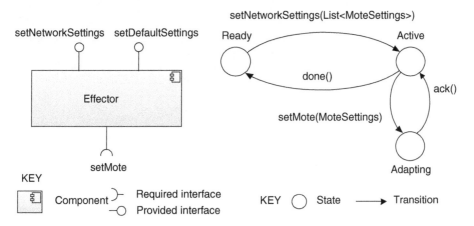

Figure 6.2 Illustration of a simple functional model for DeltaIoT.

models of the quality attributes that are the subject of adaptation and directly relate to the adaptation goals. These quality models are used to analyze the different adaptation options based on up-to-date knowledge. Based on this analysis and the adaptation goals, the managing system can then select the best option to adapt the managed system.

The ISO 9126/25010 standard defines a quality model as "the set of characteristics, and the relationships between them, that provides the basis for specifying quality requirements and evaluation." Hence, quality models specify the properties that are important for a system (e.g. its performance, reliability, or security) and how these properties are to be determined. Key properties of quality models are discretization, relevance, and accuracy. Discretization is important to represent modeled aspects with a continuous domain by means of discrete values allowing reasoning. Discretization also provides a means of abstraction, which is important when not all details of what is modeled are known. Relevance is obviously important as representations of the system and other aspects related to it are used for reasoning and decision-making about the system. Finally, an important aspect is accuracy as there is often only limited information, and hence predictions made based on quality models are often subject to ambiguity.

A wide variety of runtime quality models have been developed over the years. We elaborate on two prominent examples: Markov models and queuing models.

A Markov model is a stochastic model that represents the possible states of a system or an environment and the transitions between them with probabilities that the transitions will be taken. A Markov model assumes that future states depend only on the current state, not on the events that occurred before it, which is called the Markov property. The states of Markov models can be fully observable or partially hidden. Hidden Markov models allow a sequence of unknown (hidden) variables to be predicted from a set of observed variables. Markov models can, for example, be used to model the reliability of a system, where the reliability can be evaluated as the probability that the system arrives in a failure state.

Example Figure 6.3 shows a simple Markov model that represents the communication of packets over a single link in the DeltaIoT network.

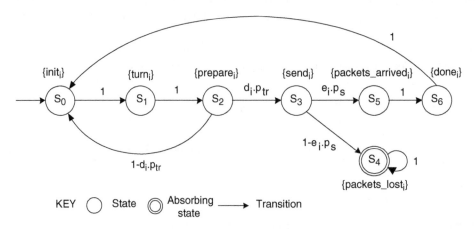

Figure 6.3 Illustration of a simple Markov model for DeltaIoT.

The model is triggered when a mote receives its turn in the time schedule to communicate packets over a link. Whether the mote will actually send packets or not is determined by the probability $d_i.p_{tr}$. The value of this probability is determined by two factors: on the one hand, the distribution of packets sent by the mote over the link (d_i), and on the other hand the uncertainty of fluctuating traffic load produced by the mote (p_{tr}). The distribution of packets is a parameter that can be set by the system. The value of the traffic load needs to be updated by the managing system according to observations of the actual conditions. When the mote is not sending packets, it returns to the initialization state, where it waits for its next turn. Otherwise, the mote sends packets. The packets will either arrive at the parent mote or will get lost (absorbing state). Whether the packets will arrive or get lost is determined by the probability $e_i.p_s$. The value of this probability is also determined by two factors: a value that depends on the energy setting of the mote used to communicate packets over the link (e_i), and the uncertainty of influence of the network interference and noise on the communication along the link (p_s). Here too, the value of the energy setting is a parameter that can be set by the system, while the value of the network interference along the link needs to be updated according to observations. The communication slot ends when the packets have arrived. To represent the communication of the overall network, the Markov models for all links are combined. The resulting model allows the expected packet loss for different configurations to be determined that are defined by different settings for the distribution of packets (d_i) and energy settings (e_i) of the links in the network. Such analysis typically requires proper tool support, such as PRISM, which is a well-known tool for the analysis of Markov models.

A queuing model represents a set of servers that perform tasks for clients. When client tasks arrive and they find all servers busy, they will join one of the server queues from which the respective server will pick jobs. A simple example of a daily life queuing system is the cashiers in a supermarket. Queuing models can be used to determine the performance of queuing systems. For instance, a queuing model can be used to measure the utilization of the server, to predict the average waiting time for a task to be processed, or to determine the expected latency of a network. The analysis of queuing models can be done by dedicated tools that employ analysis techniques derived from queuing theory. Since these techniques are known to be complex, in practice the analysis is often based on running simulations of the queuing model.

Example Figure 6.4 shows a simple queuing model that models the processing of packets by a subset of motes of the network in DeltaIoT.

The model represents each mote as a server that processes packets that it takes from a queue with a given size. The model abstracts away from the internal queues of a mote. Packets can originate from a child mote or they can be generated by the mote itself based on data obtained from local sensors. Packets are communicated in relative order as indicated by the number on the arrival links. When a mote receives a slot to communicate packets over a link, it picks the packets from its queue and sends these based on a particular strategy – for instance first-in-first-out – until the slot ends. The remaining packets stay in the queue until the next slot. The arrival of packets along a link is modeled as a process with a particular probability distribution, which is determined by the traffic produced by the mote and its children. The model allows various analyses to be performed. For instance, the system may

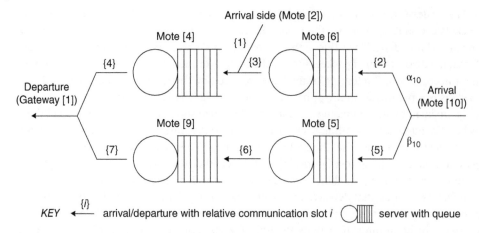

Figure 6.4 Illustration of a simple queuing model for DeltaIoT (flow from left to right).

simulate the model and observe the effect on the latency of the subnet when the distribution of packets sent by Mote [10] is changed (α_{10} and β_{10}). Such analysis may be used by a managing system to adapt the settings of the network to avoid a violation of the latency goal. Alternatively, the effect of the traffic injected into the subnet by Mote [2] *(Arrival side)* may be evaluated. This extra traffic may cause queue loss at Mote [4], i.e. an overflow of the queue of Mote [4]. To avoid such an overflow and guarantee the packet loss goal, a bound may be set for the traffic injected into the subnet by Mote [2] when adaptation options are evaluated.

A Note on Analysis based on Simulation To obtain trustworthy results when simulation is used for the analysis of a probabilistic runtime model, it is important to note that the managing system needs to apply a proper statistical treatment to the results based on the characteristics of the input data and the properties of the model at hand. Theoretical bounds have been identified on the number of simulation runs that are required to ensure a particular level of accuracy and confidence for the simulation results of particular types of probabilistic models. Yet in practice, more effective methods may be applied. In general, the design of runtime analysis based on simulation is a non-trivial task.

6.4.4 Formal versus Informal

A final dimension we highlight is the distinction between formal and informal representations used in runtime models. This distinction is primarily a matter of rigor versus practicality.

Formal runtime models are based on mathematical principles, usually discrete mathematics. The modeling primitives have a well-formed syntax and semantics. Since formal models are well-defined, they can be precisely communicated. Formal runtime models support automated reasoning about the state and behavior of a self-adaptive system. Automated reasoning is usually required for systems with strict adaptation goals that need to be guaranteed. The analysis of formal models will yield results that can be replicated. In principle, any relevant aspect of a self-adaptive system can be represented in a formal model. This ranges from a formal specification of the adaptation goals of a system in some logic to a formal model of a workflow of the system that specifies the sequence of activities performed by the managed system, for instance as an automaton model. In general, no formal model describing the world is completely correct, and it is more difficult to adequately capture concepts in the domain using formal models.

Informal runtime models, on the other hand, use modeling primitives that are based on programming models without a formal basis, domain abstractions, and stakeholder advice. Informal models have no complete mathematical underpinning and hence are less precise. As such, some aspects of informal models may be subject to interpretation. However, informal runtime models may be used for practical reasons. In particular, it may be impossible or impracticable to model a system or aspects related to it in a formal way using mathematical techniques. This may be because no methods are available to adequately specify the system, or because the resources required for using formal methods are not justifiable for the self-adaptation problem at hand. In these cases, an informal modeling approach can be used.

Example   Figure 6.5 illustrates a formal runtime model that can be used to predict the failure rate in the health assistance system that we introduced in Chapter 1.

The model is specified as a timed automaton. A timed automaton is a finite state machine that can be extended with real-valued clocks that progress synchronously. Timed automata can be combined in a network. The state of a network of timed automata is defined by the state of all automata, the clock values, and the values of the ordinary variables. Only one state per automaton is active at a time. Automata can synchronize through channels. Depending on the type of the channel (binary or broadcast), a sender $x!$ can synchronize with one or more receivers $x?$ through a signal x. An edge can be annotated with a *guard* that expresses a condition that must be satisfied for the edge to be taken and functions that are executed when the transition along the edge is taken. States marked with a C are committed states. No delay is allowed to occur in a committed state. States marked with a U are urgent states. A process needs to leave an urgent state with priority over regular states without delay.

The runtime model of Figure 6.5 shows only the automaton that computes the expected failure rate of a particular service configuration of the health assistance system. This automaton, which represents the behavior of a workflow of the system, has a set of parameters that can be set dynamically during analysis. The first set of parameters corresponds to the service instances of each service type (these parameters are not shown in Figure 6.5). The permutation of all available service instances per service type determines the set of adaptation options. Each selected service will be characterized by a particular failure rate, which is tracked and maintained by the monitor. Furthermore, the values of other parameters that represent uncertainties, which are also tracked by the monitor, are set; e.g. the value of *p_CHANGE_MEDICATION* represents the probability that the drug service will be invoked after the analysis of the vital parameters by the medical service. Once the parameters are set, the estimated failure rate of the selected service combination is determined by running the model simulation a number of times.

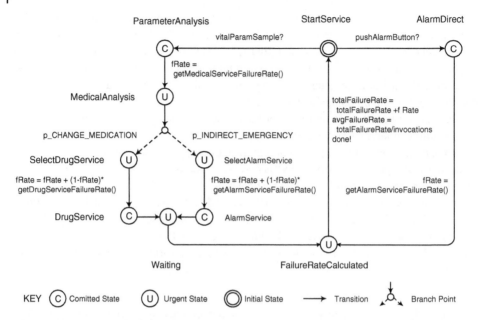

Figure 6.5 Example of a formal model to predict the failure rate of a service configuration in the health assistance system (introduced in Chapter 1).

The automaton, which is used by the managing system, is activated from the initial state *StartService* either by a message from the user who pushes the alarm service button or by a message from the device that takes a sample of the vital parameters. When the user pushes the button, the alarm service will be activated, but it may fail with a failure rate that depends on the alarm service selected for the invocation. On the other hand, if a sample is taken, this sample is sent to the medical service that will perform the analysis. This invocation may fail, depending on the failure rate of the medical analysis service. Depending on the analysis result represented by the probabilities *p_CHANGE_MEDICATION* and *p_INDIRECT_EMERGENCY*, either the drug service or the alarm service is invoked at the branch point. The invocation of these service instances may fail as well, depending on the failure rates of the respective service instances. When the path of service invocations in the workflow completes (both for the case when the user directly pushed the alarm button or for the case that a sample with vital parameters was taken), the total failure rate is computed as well as the average failure rate per invocation. This process is repeated for a predefined number of service invocations to ensure accurate and confident results.

This example illustrates how a formal runtime model can be used to predict a particular quality property for a given adaptation option with a required level of accuracy and confidence. By ranking the adaptation options based on the predictions of their qualities and comparing them with the required adaptation goals, the self-adaptive system will be able to select the best option for adaptation.

6.5 Principal Strategies for Using Runtime Models

We direct our focus now from *what* a runtime model is to *how* a runtime model can be used in the realization of self-adaptation. To that end, we present three common strategies that have been applied in the use of runtime models. Figure 6.6 gives a high-level overview of the three strategies.

In strategy (a), MAPE components share a common set of runtime models. This strategy is the most common approach to the use of runtime models in self-adaptive systems. In strategy (b), MAPE components exchange runtime models; the components only exchange the models they need to realize their functions. Finally, in strategy (c), MAPE models share a common set of runtime models. Strategies (b) and (c) are less frequently used in self-adaptive systems. Independently of the strategy, the runtime models map to the knowledge in the reference model of a managing system, see Figure 4.7. Strategies differ in the benefits they offer but also in the consequences they impose. Note that in practice, a mix of strategies may be applied. The following sections elaborate on the three principal strategies.

6.5.1 MAPE Components Share K Models

In the first strategy, the runtime models are stored in a shared knowledge repository that is accessible by the MAPE components. We illustrate the first strategy for a concrete realization as shown in Figure 6.7. This realization is based on MARTAS, short for Models At RunTime And Statistical techniques, a typical approach to realizing the first strategy.

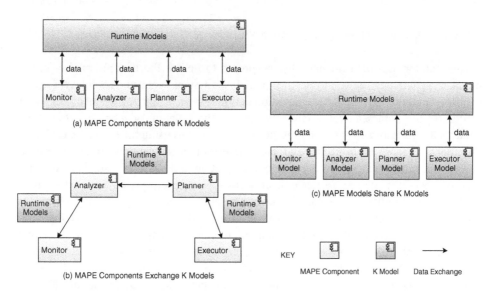

Figure 6.6   High-level overview of the three strategies for using runtime models in self-adaptation.

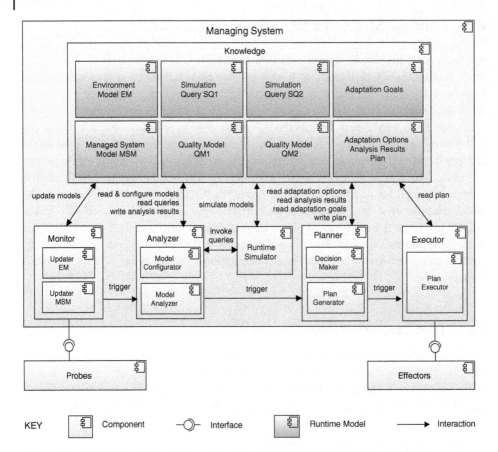

Figure 6.7 Illustration of the strategy where MAPE components share runtime models.

The monitor uses data provided by the probes to update the runtime models of the managed system and the environment in which the system operates. The analyzer uses a runtime simulator to analyze the adaptation options. In particular, the model configurator configures the parameterized quality models for the different adaptation options and instantiates the parameters that represent the uncertainties using the current knowledge. The model analyzer then invokes simulation queries that the runtime simulator uses to analyze the quality models for the different adaptation options. This results in estimates for the different quality properties of each adaptation option. The analyzer writes these analysis results to the knowledge repository. The planner then reads the analysis results for the different adaptation options. The decision maker uses this data together with the adaptation goals to select the best adaptation option (or applies a fail-safe strategy if needed). The plan generator then composes a plan to adapt the system from its current configuration to the configuration of the selected adaptation option and writes this plan to the knowledge repository. Finally, the executor reads the plan and executes it via the effectors.

This example shows how the MAPE components coordinate their work by reading and writing data to the runtime models. An important benefit of the first strategy is that all knowledge is accessible by the MAPE components. However, if MAPE components can operate in parallel and access the runtime models concurrently, a synchronization mechanism is required to avoid race conditions that may corrupt the shared data. Alternatively, the rights of the different components to access the repository may be restricted through dedicated interfaces with specific rights.

6.5.2 MAPE Components Exchange K Models

In the second strategy, the MAPE components exchange runtime models that they can reason upon and manipulate. Eventually, the flow of activities performed on the runtime models results in an adaptation of the managed system such that the adaptation goals are achieved.

Figure 6.8 shows an example of a concrete realization of the second strategy. This example is based on the principles of DiVA, short for Dynamic Variability in complex Adaptive systems, which is a well-known approach to realizing this strategy.

The self-adaptive system is structured in three layers. The bottom layer, *Business Application*, contains the application components and is equipped with sensors that track runtime events from the application and its environment, and factories that enable instantiating new business component instances. The top layer, *Online Model Space*, consists of the adaptation components that manipulate models. The middle layer, *Causal Connection*, implements a weak causal connection between the online model space and the business application.

The bottom layer of the self-adaptive system corresponds to the managed system in the conceptual model of a self-adaptive system. The top two layers correspond to the managing system, which consists of five components that interact by exchanging four types of runtime models.

6.5.2.1 Runtime Models

The *feature model* describes the variability of the system. The model specifies mandatory, optional, and alternative features, as well as constraints among the features. Examples of such constraints are: one feature requires another, and a feature excludes another feature. Features refer to architectural fragments that realize the functionalities of the features. To that end, the feature model uses a specific naming convention. The granularity of the architectural fragments determine the resolution of the adaptations supported in the system.

The *context model* specifies relevant attributes and processes of the environment in which the system executes. The knowledge of the context model is kept up-to-date at runtime using sensor data. The updates of the context model may trigger adaptations of the business application, as we explain below.

The *reasoning model* associates sets of features with a particular context. A reasoning model can be instantiated as a set of event-condition-action rules. An event specifies a signal that triggers the invocation of a rule, e.g. the signal that a particular service has failed. The condition part provides a logical expression to test whether the rule applies or not, e.g. to check whether the functionality of the failed service is required in the current context. The action part consists of update actions that are invoked if the rule applies, e.g. unbind the failed service and bind a new alternative service.

Finally, the *architecture model* specifies the component composition of the business application. The specification requires an architecture description language that supports

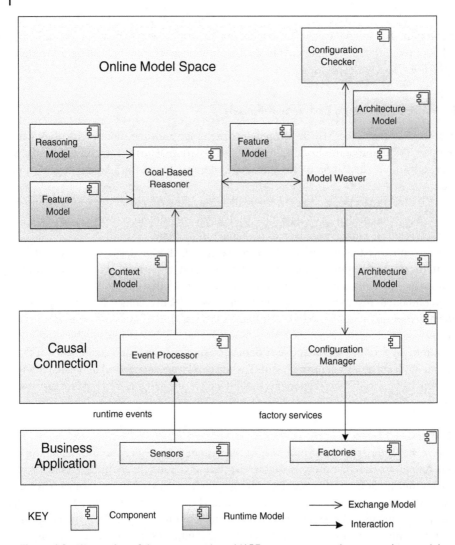

Figure 6.8 Illustration of the strategy where MAPE components exchange runtime models.

dynamic reconfigurations of components. Designers can either use an off-the-shelf language or they can specify an ad-hoc meta-model tailored to the needs of the problem at hand. Such a tailored language reduces the memory overhead at runtime. The architecture model refines each leaf feature of the feature model into a concrete architectural fragment.

6.5.2.2 Components of the Managing System

The *Event Processor* observes runtime events from the running system and its context. When relevant events occur, the event processor updates the context model of the system. The update process can be as simple as changing the value of a variable of the context model up to complex event processing, such as aggregating data, removing noise, and online learning.

When the *Goal-Based Reasoner* receives an updated context model from the event processor, it uses the feature model and reasoning model to derive a specific feature model that is aligned with the current context. This specific feature model contains all the mandatory features that the system requires and a selection of optional features that the system needs in the current context.

When the *Model Weaver* receives the specific feature model from the goal-based reasoner, it uses the feature model to compose an updated architecture model of the system. This architectural model specifies a new configuration for the system aligned with the current context.

Before the system is updated, the model weaver sends the updated architecture model to the *Configuration Checker*. This component evaluates the consistency of the new configuration of the running system. This evaluation includes checks of generic invariants of the configuration and possibly user-defined application-specific invariants. If the configuration is valid, the model weaver sends the architecture model to the configuration manager. Otherwise, the model weaver triggers the goal-based reasoner to compute an alternative configuration.

The *Configuration Manager* will reconfigure the architecture of the business application according to the new architecture model it received from the model weaver. Based on the delta between the configuration of the running system and the new configuration, the configuration manager will deduce a safe sequence of reconfiguration actions. This sequence can then be encoded in a script that the underlying infrastructure can execute to adapt the business application. The configuration actions, which are enacted by the script, include removing, instantiating, and adding components and bindings.

This example illustrates how MAPE components can exchange runtime models to realize self-adaptation. An important benefit of the second strategy is the modularity provided by the exchange of runtime models, which allows the adaptation logic to be easily evolved by adding or removing invariants, constraints, and rules. However, the second strategy only works well if components can exchange the runtime models in an efficient manner. In the example this is achieved by considering only coarse-grained adaptations of the system requiring relatively small models.

6.5.3 MAPE Models Share K Models

The third strategy extends the application of runtime models to the adaptation logic itself. In particular, in this strategy runtime models of the knowledge are shared by MAPE functions that are specified as runtime models. The MAPE models are directly executed at runtime to realize self-adaptation.

Figure 6.9 shows an example architecture of a concrete realization of the third strategy. This example is based on the principles of EUREMA, short for Executable Runtime Megamodels, which is a pioneering approach that realizes the third strategy.

Reflection Models represent the relevant architectural aspects of the managed system and its environment. These models are shared among all MAPE models. The reflection models are kept up-to-date by the *Monitor Model* using a set of *Monitoring Models*. Monitoring models map system-level observations to the required level of abstraction used in the reflection models. The *Analyzer Model* performs an analysis of the reflection models to identify

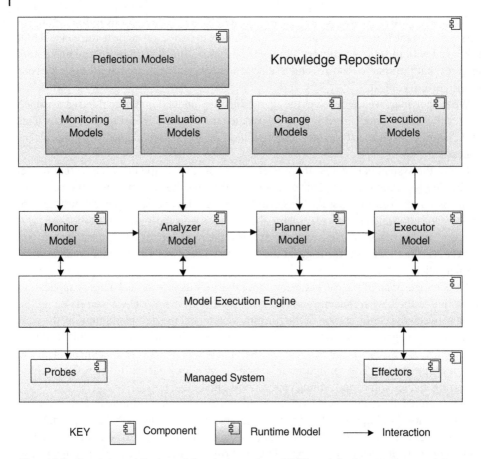

KEY   Component   Runtime Model   ———▶ Interaction

Figure 6.9 Example architecture of the strategy where MAPE models share runtime models.

the need for adaptation. The need for adaptation is determined using *Evaluation Models*, which define allowed configurations and the required utility of the managed system – for instance, using constraints on reflection models and thresholds on quality properties of the system. If there is a need for adaptation, the *Planner Model* will devise a plan that prescribes the required changes of the system reflected in the reflection models. Planning uses *Change Models* that define the space of allowed variability of the system. Planning explores this space to find an appropriate configuration to adapt the managed system to based on a representation of the adaptation goals. Finally, the *Executor Model* enacts the adaptation plan using a set of *Execution Models* that map model-level steps of the plan to concrete system-level adaptation actions. The runtime feedback loop models are directly executed by a *Model Execution Engine* to realize the adaptations of the managed system. This engine is also responsible for the synchronization of the interaction of the feedback loop models with probes and effectors.

Figure 6.10 illustrates the workflow of the analysis and plan stage for handling failures. The analyzer applies a set of failure analysis rules to the architecture model of the managed system to detect failures of the system. If no failures are detected, no further action

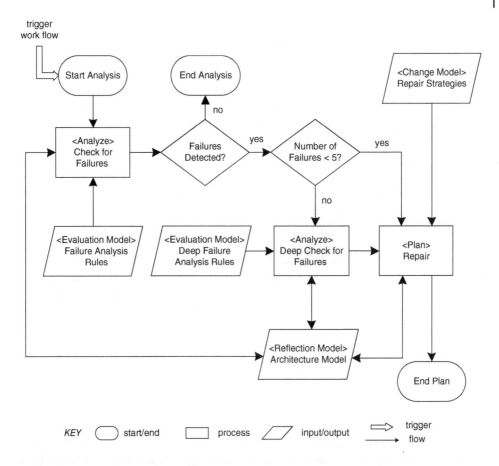

trigger
work flow

Figure 6.10 Illustration of the strategy where MAPE models share runtime models.

is required. If less than a given number of failures are detected (five in the example scenario), the planner is triggered. If more failures are detected, a second in depth analysis is
performed based on a set of deep failure analysis rules, before the planner is triggered. To
mitigate the identified failures, the planner composes a plan using a set of repair strategies.
The selected strategy uses the current architectural model to prescribe a reconfiguration of
the software.

This example shows how the principles of runtime models can be extended to the MAPE
functions themselves. An important benefit of the third strategy is that this model-centric
approach paves the way for supporting dynamic evolution of the runtime models, which
is a key aspect of goal management (cf. the three layer model for self-adaptive systems in
Section 5.2). However, applying the third strategy comes with a number of consequences,
including potential limitations of the expressiveness of the high-level language used to specify models and the need for a trusted model execution engine to execute the MAPE models
specified in the modeling language.

6.6   Summary

Runtime models support engineers with the concrete design of self-adaptive systems. Runtime models leverage the principles of model-driven engineering into the runtime.

A runtime model of a self-adaptive system is a first-class runtime abstraction of that system or any aspect related to it that is used for the purpose of realizing self-adaptation.

Causality ensures consistency between a runtime model and the represented system or aspect of it. In self-adaptation, it is hard to realize a strict causal connection for all types of runtime models. Hence, in self-adaptive systems we use weak causal connection, which allows a discrepancy between the state of a runtime model and the state of the system or aspect of the system it represents. This discrepancy should not invalidate the decision making of the managing system.

Various arguments underpin the motivation for the use of runtime models. The abstraction brought by runtime models enables us to focus on the facets that are relevant to dynamic change in self-adaptation. Runtime models enable separation of concerns, each model capturing a different aspect of the system or its environment. Runtime models enable the use of different types of reasoning, supporting the decision-making of adaptation. Since runtime models are abstractions, they can leverage the involvement of human stakeholders in different aspects of adaptation. Well-modularized runtime models facilitate on-the-fly evolution of the feedback loop system.

Runtime models can be classified according to several dimensions. Models can either represent structural aspects of the system or behavioral aspects. They can be declarative – representing knowledge about what is needed or holds – or procedural – representing how something can or needs to be done.

Runtime models can focus either on functional aspects of the system or on qualitative aspects. Runtime quality models allow a self-adaptive system to analyze changing conditions and make adaptation decisions. Example model types to represent system qualities are Markov models and queuing models.

Formal runtime models have a well-defined syntax and semantics, based on mathematical principles. These models allow automatic reasoning. When it is not possible or not preferable to use formal runtime models, informal runtime models may be applied. While being practical, the lack of precision of such models may lead to varying interpretations.

Three principal strategies can be used to apply runtime models in self-adaptive systems. In the first strategy, MAPE components share knowledge models. A benefit of this common approach is the shared access to the runtime models by all the feedback loop components. However, sharing knowledge concurrently may require synchronization schemes to avoid race conditions.

In the second strategy, MAPE components exchange runtime models to realize self-adaptation. This strategy enables the adaptation logic to be easily modified. However, this strategy requires that the models can be exchanged efficiently, which may have an impact on the scale of the system that can be supported or granularity of the adaptations that can be applied.

In the third strategy, the MAPE functions are realized as runtime models themselves, similar to the knowledge models they share. The MAPE models are directly executed at runtime by an execution engine to realize adaptation. By representing all the adaptation

Table 6.1   Key insights of Wave III: Runtime Models.

• Different runtime models enable managing the complexity that arises from the concrete design of self-adaptive systems.

• A runtime model of a self-adaptive system is a first-class runtime abstraction of that system or any aspect related to it that is used for the purpose of realizing self-adaptation.

• In self-adaptation we usually use weak causal connection, which tolerates a discrepancy between the state of a runtime model and the state of the self-adaptive system represented, or aspects of it.

• Four key dimensions of runtime models are: structural versus behavioural, procedural versus declarative, functional versus qualitative, and formal versus informal.

• Three principal strategies to applying runtime models are: MAPE components share knowledge models, MAPE components exchange runtime models, and MAPE models share runtime models.

---

logic as runtime models, this strategy paves the way for supporting runtime evolution of the managing system. However, the third strategy has consequences as well, including the need for a trusted model execution engine and possible limitations in terms of the expressiveness of the modeling language.

Table 6.1 summarizes the key insights of Wave III.

6.7   Exercises

6.1   Descriptive vs. prescriptive runtime models: level H

In this chapter, four dimensions of runtime models are presented. An additional dimension could be descriptive vs. prescriptive models. The former tries to explain how things are, the latter, how they should be. The former is what science and reverse engineering does, the latter what engineering does. Elaborate on descriptive vs. prescriptive runtime models and illustrate with examples. Evaluate why this extra dimension could be useful or not.

6.2   Runtime models of MoRE: level D

MoRE, short for Model-Based Reconfiguration Engine, offers a platform for self-adaptation that is based on the principles of dynamic software product lines. MoRE is able to determine how a system should evolve based on changes in its context. Central to MoRE is a feature model that specifies the functionality of a system at a coarse-grained level. Features are hierarchically linked in a tree-based structure through relationships, such as optional, mandatory, single-choice, and multiple-choice. Each feature, which can be active or inactive, is implemented by a specific set of architectural elements of the system. Whether a feature should be active or inactive is determined by the actual conditions of the context in which the system operates. To that end, MoRE introduces the notion of resolution that represents the set of required feature changes triggered by a context condition. The task of self-adaptation is to activate or deactivate features dynamically based on the observed context conditions. MoRE has been applied to a smart home case, with a focus on self-healing and self-configuration. Study the MoRE approach and map its architecture to a classic MAPE-based feedback loop. Highlight the core runtime models it uses. Discuss the responsibilities of the different elements. Instantiate the architecture for a smart home setting. Illustrate the adaptation workflow for a concrete self-healing and self-configuring scenario. For the MoRE paper, see [47].

6.3 Integrating self-healing and self-optimization in mRUBIS: level W

Setting. mRUBIS is an artifact that simulates a marketplace on which users can auction and sell items. The artifact supports model-based adaptation with a focus on self-healing and self-optimization. The artifact comes with a runtime model that is kept updated during execution. This runtime model can be used when developing a managing system to realize self-adaptation. mRUBiS supports the injection of issues into the managed system, which are represented to the adaptation logic via the runtime model. These issues then need to be handled through self-adaptation.

Task. Download the mRUBIS artifact and install the software. Experiment with the example applications that come with the artifact. Focus on the instances that support self-healing and self-optimisation based on a MAPE structure. Specify the design of the feedback loop and the runtime models for the instance with self-healing. Explain the responsibilities of the elements and models, and explain how they work together. Repeat for the instance with self-optimisation. Experiment with both instances, and study the effects of different parameter settings of the utility functions used by the respective managing systems. Now design a new self-adaptive solution that integrates both self-healing and self-optimization. Implement your solution and test it using the artifact. Critically assess your solution.

Additional material. mRUBIS can be downloaded from the artifact website [194]. For background information about the example cases, see [193].

6.4 Adaptation strategy with cost for Lotus@Runtime: level W

Setting. Lotus@Runtime is a tool that uses a runtime model to identify the need for self-adaptation. The tool enables a designer to specify a model of the managed system as a Probabilistic Labeled Transition System (PLTS). This model can then be used by a monitoring and analysis infrastructure that is provided by the tool. At runtime, the monitoring part of the infrastructure tracks execution traces of the system, and based on the occurrences of different system events, the infrastructure updates the probabilities of the corresponding transitions on the system model. Then, the analysis part of the infrastructure can verify the updated probabilistic model against a set of reachability properties. If a property is violated, the planner and executor of the managing system can be notified to take appropriate action.

Task. Download and install the Lotus@Runtime tool. We focus on the Travel Planner application, which is a service-based system that provides services to search for flights, tourist attractions, accommodation arrangements, and rental cars or bicycles. Specify the design of the self-adaptive system with the feedback loop and the runtime models, and explain the responsibilities of the different elements. Explain the adaptation workflow. Run the Travel Planner application and experiment with the default service selection strategies, taking into account the predefined requirements for failure rates of the different types of services. Compare the results with the evaluation results provided in the artifact package. Now add a requirement that minimizes the cost of service invocations, and design a novel adaptation strategy that takes into account both reliability and cost. Implement your solution, evaluate it, and critically assess it. **Additional material.** The Lotus@Runtime tool can be downloaded from the artifact website [12]. The example applications can be downloaded here [13]. For the evaluation results of the Travel Planner application with default settings, see [11]. This article also provides additional information about the tool.

6.5 Quality models for simple IoT network: level D

Setting. Consider the simple IoT network (similar to DeltaIoT) as shown in Figure 6.11.

The network consists of five motes that collect local information from the environment via sensors. The communication in the network is time-synchronized. Each cycle consists of eight time slots. During the first six slots, the motes can communicate the sensor data to the gateway in the order indicated by the communication slots shown in Figure 6.11. During the last two slots, the gateway can directly communicate with the motes (these communication slots are not shown in the figure). The systems should achieve the following goals:

AG1: The average packet loss over 10 cycles should be less than 10%;

AG2: The energy consumption over 10 cycles should be minimized.

In each cycle, the motes generate messages as shown in Table 6.2.

The number of messages generated by Mote [2] switches every three cycles between zero and five, as shown in Figure 6.12.

This pattern represents domain knowledge. However, the phase of this pattern is unknown; this is an uncertainty.

The motes are equipped with a buffer to store 10 messages. In each time slot, a mote can transmit 10 messages. Messages that arrive when the buffer is full are lost.

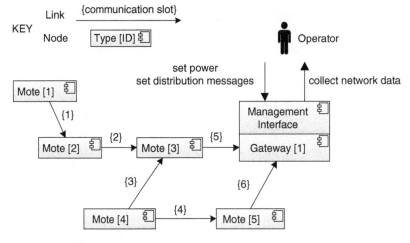

Figure 6.11 Simple IoT network.

Table 6.2 Messages generated by the motes in the simple IoT system.

Mote	Messages generated
Mote [1]	8
Mote [2]	0 or 5
Mote [3]	3
Mote [4]	5
Mote [5]	7

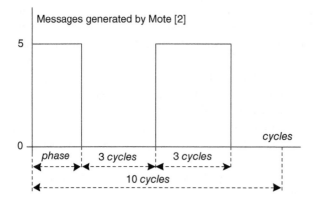

Figure 6.12 Example of messages generated by Mote [2].

At the gateway, a management interface of the IoT network is available that provides two options to change the network settings:

- The power settings of the motes can be set to 5 or 10. The setting applies to all motes.
- Mote [4] can either sent 100 % of the messages in its buffer either to Mote [3] or to Mote [5], or alternatively, the messages are evenly distributed to both parent motes.

The basic signal-to-noise ratios (SNR in dB) along the links between the motes for different power settings are shown in Table 6.3.

The SNR levels shown in Table 6.3 are subject to uncertainty caused by interference in the environment. In particular, the SNR values are decreased either with a value of -2 dB along all the links or with a value of -5 dB. The level of interference in the environment changes every three cycles, following a similar pattern to the messages generated by Mote [2] as shown in Figure 6.12. The interference pattern represents additional domain knowledge, yet the phase of the pattern is unknown and not necessarily in synchrony with the pattern of how Mote [2] generates messages.

The packet loss for a link depends on the value of the SNR. The following formula can be used to determine the expected packet loss for a given SNR value and the number

Table 6.3 SNR (dB) along the links in the simple IoT system.

Link	SNR (power setting 5)	SNR (power setting 10)
Mote [1] - Mote [2]	-1	8
Mote [2] - Mote [3]	0	5
Mote [4] - Mote [3]	2	4
Mote [4] - Mote [5]	2	5
Mote [3] - Gateway [1]	-2	6
Mote [5] - Gateway [1]	4	6

of messages to be sent over that link

$$Packet\_Loss = \begin{cases} 0 \text{ if } SNR \leq 0 \\ \frac{x \times (-SNR)}{20} \text{ } otherwise \end{cases} \qquad (6.1)$$

where x is the number of messages to be sent over the link, and SNR is the actual value of the signal-to-noise ratio.

Motes consume energy when sending and receiving messages. To calculate the energy consumed in sending messages, the following formula can be used

$$Send\_Energy\_Consumption = \begin{cases} x \times 1.158 \times 26.1 \text{ } for \text{ } power \text{ } setting \text{ } 5 \\ x \times 1.158 \times 32.4 \text{ } for \text{ } power \text{ } setting \text{ } 10 \end{cases} \qquad (6.2)$$

where x is the number of messages sent, 1.158 is an empirically determined factor that depends on the transmission time that is required to send a message, and 26.1 and 32.4 are empirically determined factors that depend on the energy required to send a message with the given power settings.

To calculate the energy consumed for receiving messages, the following formula can be used

$$Receive\_Energy\_Consumption = x \times 1.158 \times 14.2 \qquad (6.3)$$

where x is the number of messages received, 1.158 is a factor that refers to the transmission time, and 14.2 is a factor that refers to the energy used to receive a message.

Task. Design a timed automata model using the Uppaal tool that allows the packet loss for a given setting of the IoT to be predicted. Define a test automaton, and test the model using the simulation mode of Uppaal. Change the setting and observe the effects on the packet loss. Repeat the task for energy consumption. Assess your solution.

Additional material. The Uppaal tool can be downloaded at http://www.uppaal .org/. This website provides a number of tutorials to get familiar with the tool.

6.8 Bibliographic Notes

G. Blair et al. pioneered the notion of "model at runtime" [31], which the authors defined as "a causally connected self-representation of the associated system that emphasizes the structure, behavior, or goals of the system from a problem space perspective." The authors also identified key dimensions of runtime models. The basic principles of runtime models and the dimensions of runtime models described in this chapter are based on these insights. For an interesting collection of papers on models at runtime, see [23]. N. Bencomo et al. performed a systematic literature review on research in models at runtime in [24].

P. Maes elaborated on causality in computational reflection [130]. An interesting relaxation of strict causality is so called probabilistic causality, which is informally defined as "*A* probabilistically causes *B*." A gentle philosophical discussion of probabilistic causality can be found in [139].

M. Autili et al. present an approach to the automated synthesis of connectors from automata-based models [9]. This approach forms a basis for on-the-fly generation of connectors that allow components with similar but not identical interfaces to interact with one another.

For the ISO 9126/25010 standard on quality models, see [105]. An introduction to Markov chains can be found in [157]. For a gentle introduction to Markov models and hidden Markov models, see [77]. For an introduction to queuing networks, see [181]. C. Baier and J. Katoen provide a comprehensive introduction to automata and the foundations of model checking [10]. For more information on timed automata with a discussion of the semantics of committed and urgent states, see [57].

MARTAS, the approach that we used to illustrate the strategy MAPE components that share K models, is introduced in [211]. DiVA, which applies the strategy MAPE components that exchange K models is introduced in [145]. EUREMA, the approach we used to illustrate the strategy MAPE models that share K models, is described in [195].

7

Wave IV: Requirements-driven Adaptation

The fourth wave focuses on the requirements of self-adaptive systems. The first wave identified four essential types of maintenance tasks that can be automated using self-adaptation: self-optimization, self-healing, self-protection, and self-configuration. These types of maintenance tasks imply high-level objectives of self-adaptation. The third wave emphasized the role of runtime models as a means to manage the complexity of the concrete design and implementation of self-adaptive systems that automate such maintenance tasks. The objectives of a self-adaptive system are expressed, explicitly or implicitly, as adaptation goals. While adaptation goals capture the concerns of the stakeholders of a self-adaptive system, their purpose is mainly operational: adaptation goals are in essence machine-readable specifications that are operationalized by a feedback loop to drive the adaptation of the managed system.

The fourth wave zooms in on the requirements of self-adaptive systems from a stakeholder perspective. The drivers for this wave were the insights obtained from the first and third wave, in particular the comprehension that when designing feedback loops, it is essential to understand the requirements problem they intend to solve. Without this understanding, self-adaptive systems are constructed in an ad-hoc manner. Hence, handling the requirements of self-adaptive systems as first-class citizens is an important facet of any systematic engineering process.

In general, requirements engineering consists of various activities, including elicitation, analysis, specification, operationalization (i.e. linking requirements with system elements), and maintenance of requirements. The concrete realization of these activities depends on both the characteristics of the system that is developed and the practices and experiences of the teams involved. Here we focus on the specification and operationalization of requirements for self-adaptive systems.

In this chapter, we study three important approaches for requirements engineering of self-adaptive systems. The first approach relies on relaxing requirements for self-adaptive systems to mitigate uncertainties. The second approach models the requirements for adaptive behavior as meta-requirements, i.e. requirements about the regular requirements of the system. Finally, the third approach looks at requirements of the functional behavior of feedback loops.

An Introduction to Self-Adaptive Systems: A Contemporary Software Engineering Perspective,
First Edition. Danny Weyns.
© 2021 John Wiley & Sons Ltd. Published 2021 by John Wiley & Sons Ltd.

LEARNING OUTCOMES

- To explain and motivate the need for requirements engineering of self-adaptation.
- To explain and illustrate the relaxation of requirements for self-adaptation.
- To explain and illustrate meta-requirements for self-adaptation.
- To motivate, explain, and illustrate functional requirements of feedback loops.
- To apply the different approaches for requirements engineering of self-adaptation to examples.

7.1 Relaxing Requirements for Self-Adaptation

In general, requirements refer to the functions that a system-to-be should achieve (*what* should be realized) and the qualities the system should have (*how* the functions should be realized). In addition, a rationale may be provided by the stakeholders for the requirements of the system-to-be (*why* should the requirements be realized). Over the years, numerous languages and notations have been developed to specify requirements. Well-known examples are use cases and user stories to specify functional requirements, and structured natural language, quality attribute scenarios, and goal models to specify quality requirements.

Compared to traditional approaches for the specification of system requirements, languages and notations for specifying requirements of self-adaptive systems need to explicitly address the uncertainties that are an inherent characteristic of these systems. In this section, we zoom in on an approach based on structured natural language that allows the requirements of adaptive behavior to be relaxed to mitigate uncertainty.

7.1.1 Specification Language to Relax Requirements

Structured natural languages used for requirements specifications prescribe behaviors of the system using modal verbs, such as *SHALL* and *WILL*. These verbs define the functionalities that a software system must always provide. Due to uncertainties, it may not be feasible for a self-adaptive system to achieve such strictly defined requirements at all times.

One approach to tackling this problem is by tolerating a relaxation in the non-critical requirements. This section describes a structural language with a notation that enables specifying such relaxed requirements. In particular, the language allows specifying requirements a system could temporarily relax under certain conditions, besides requirements that should never change (i.e. invariants). The language is based on RELAX, a well-known specification language for requirements of self-adaptive systems that provides explicit support for expressing uncertainty.

7.1.1.1 Language Operators for Handling Uncertainty

The language supports two types of uncertainty: changing conditions of the execution environment that may have been difficult to predict (environmental uncertainty) and changes of the system behavior at run-time in response to lack of sufficient information about the

Table 7.1 Example operators to handle uncertainty requirements of self-adaptive systems.

Operators	Explanation
Model operators	
SHALL	The requirement must hold
MAY…OR	Requirement with alternative options
Temporal operators	
AS SOON AS POSSIBLE TO	The requirement should hold as soon as possible
IN [interval]	The requirement should hold in a time interval
Ordinal operators	
AS CLOSE AS POSSIBLE TO [quantity]	Relaxation of value of countable quantity specified in a requirement
AS MANY AS POSSIBLE [quantity]	Relaxation of number of countable quantity specified in a requirement

system's intended behavior or in response to the environmental uncertainty (behavioral uncertainty).

To handle these types of uncertainties when specifying requirements, the language provides three types of operators: modal, temporal, and ordinal. Table 7.1 shows some characteristic examples of such operators.

The temporal operator *AS SOON AS POSSIBLE TO* supports uncertainty through flexibility in the time when a requirement needs to be realized. As such, this operator and the operator *IN* allow a requirement to be relaxed. For instance, consider in DeltaIoT the requirement that the packet loss shall remain below a given threshold at all times (representing an ideal situation):

> *The system SHALL keep the packet loss under a given threshold at all times.* In short: *Maintain [Packet Loss Threshold]*

In order to handle network interference peaks that may occur during short intervals, this requirement may be relaxed as follows:

> *The system SHALL keep the average packet loss under a given threshold IN periods of 12 hours.* In short: *Maintain [Average Packet Loss Threshold IN 12 hours]*

This relaxation allows a self-adaptive system to temporarily not comply with the original requirement in order to handle increasing network interference levels during short periods of time.

The ordinal operators *AS CLOSE AS POSSIBLE TO* and *AS MANY AS POSSIBLE* enable requirements that include a countable quantity in their specification to be relaxed.

Both types of relaxation operators enable flexibility to be introduced into the statements of regular requirements that are formulated with model operators such as *SHALL*. This flexibility can be supported in several dimensions, including the possible states and configurations of a system, the duration of system states, the frequency that they occur, etc. Relaxation operators define constraints on how a requirement can be relaxed at run-time. For instance, the packet loss requirement demands a required average value within periods of 12 hours.

7.1.1.2 Semantics of Language Primitives

As for any specification language, it is important to define precise semantics for the language primitives of a requirements specification language with relaxation operators. We illustrate this for two operators using fuzzy temporal logic expressions. Temporal logic allows the specification of path expressions that quantify over possible execution paths of the system (i.e. sequences of states of the system) and state expressions that quantify over states of the system within a given path. For instance *AG p* means for all paths of the system (*A*), expression *p* always (*G*) holds. Fuzzy temporal logic enhances temporal logic with fuzzy sets. Whereas the membership of a regular set is defined by a binary function (1 or 0, i.e. member or not a member), membership of a fuzzy set is defined by a membership function with real values in the interval $[0, 1]$ that define the degree of membership. Fuzzy sets enable the definition of expressions in temporal logic with uncertainties.

As an example, the semantics of two operators of the requirements specification language that can be used to relax requirements are defined as follows:

$$IN\, t\, \phi : (AFTER\, t_{start}\, \phi \wedge BEFORE\, t_{end}\, \phi)$$

where ϕ is a valid language expression, and t_{start} and t_{end} are events that denote the start and end of interval t respectively. Intuitively, this expression simply states that ϕ is true in any state in the time interval t.

$$AS\, CLOSE\, AS\, POSSIBLE\, TO\, q\, \phi : AF((\Delta(\phi) - q) \in S)$$

where q is a countable quantity, ϕ is a valid language expression, A and F are the regular "all" and "eventually" operators respectively, $\Delta(\phi)$ counts the quantifiable and compares it with q, and S is a fuzzy set with a membership function that starts with value 1 and decreases continuously. Intuitively, this expression states that there is a function Δ such that $\Delta(\phi)$ is quantifiable and $\Delta(\phi)$ is as close as possible to 0.

7.1.2 Operationalization of Relaxed Requirements

The operationalization of relaxed requirements requires the system to be enhanced with self-adaptive behavior. Central aspects to that end are: handling uncertainties, requirements reflection, and mitigation mechanisms that deal with the uncertainties.

7.1.2.1 Handing Uncertainty

Since each relaxed requirement introduces the need for self-adaptation, it is important to indicate the uncertainty factors that allow relaxation of requirements. To that end, the engineer needs to understand the sources of uncertainty and document them. Each requirement needs to be examined to determine under which uncertainty conditions it might not be possible to satisfy the requirement. If the requirement cannot be achieved under all conditions and self-adaptation is required to enable satisfaction of the requirement, then the requirement can be relaxed with the appropriate operators. In order to realize the requirement, the associated uncertainty needs to be monitored, which requires sensors that sense the sources of uncertainty and quantify them. On the other hand, requirements that a system should satisfy under all relevant conditions have to be considered as invariants and should not be relaxed. These requirements do not need self-adaptive behavior and need to be achieved by the managed system.

7.1.2.2 Requirements Reflection and Mitigation Mechanisms

The textual language for dealing with uncertainty allows requirements to be temporarily relaxed if necessary in order to mitigate uncertainties through self-adaptation. However, this language only provides a notation to specify the stakeholder requirements of a self-adaptive system. The operationalization of the requirements into a self-adaptive system requires the integration of these specifications into a unified approach that captures the requirements with their level of flexibility, the uncertainties the system is exposed to, and the adaptations that are needed to mitigate these uncertainties.

Requirements reflection applies the principles of computational reflection to the requirements of a system – that is, requirements become first-class citizens that are available at run-time. Requirements reflection enables a system to inspect its requirements and perform analysis on them to make adaptation decisions that mitigate uncertainties. One approach toward the realization of requirements reflection for self-adaptive systems is applying goal-based modeling.

Goal-based modeling offers a means to identify and represent alternatives for satisfying the overall objectives of a system. In particular, goal-based modeling enables the engineer to structure the system objectives into a directed acyclic graph of goals. The root nodes at the top of the graph represent high-level goals, while the leaves of the graph represent system requirements. Each goal in the model (except the leaf goals) is refined as a set of subgoals. The different paths in the graph show alternatives for satisfying the overall objectives of a system. These alternatives typically express tradeoffs between different quality goals, such as performance and reliability. In a goal model of a self-adaptive system, different paths can also be used to incorporate different uncertainties the system is subject to.

Example Figure 7.1 shows excerpts of two goal models for DeltaIoT. The model on the left shows the original model without self-adaptation. In the model on the right, two types of strategies are applied to mitigate uncertainties marked with a dark shaded arrow: (1) adding a subgoal and (2) relaxing a goal. These goal models rely on KAOS, short for knowledge acquisition in automated specification, a classic goal-oriented requirements specification language.

In KAOS, goals describe required properties of a system that are satisfied by so called agents. Agents can be software components or stakeholders that interact with the system. For instance, in the original goal model (Figure 7.1 left), the IoT infrastructure and an operator are two agents that contribute to maintaining the system's quality of service (by duplicating messages over links to parent motes and tuning the settings of the transmission power, respectively).

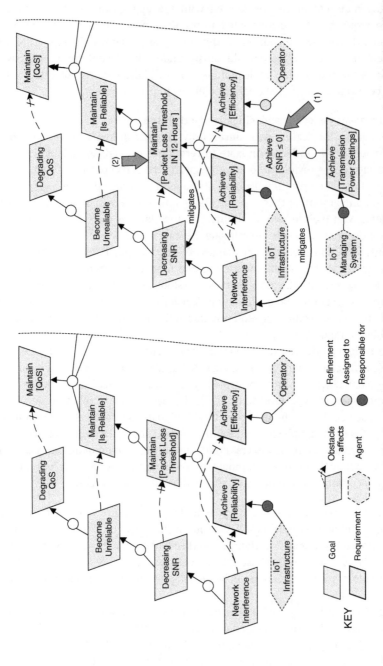

Figure 7.1 Illustration of goal models for DeltaIoT. Left: original goal model. Right: goal model with two types of uncertainty mitigations: (1) adding a subgoal; (2) relaxing a goal.

Goal models typically start with high-level stakeholder goals define concrete requirements. These goals are then refined into more concrete sub-goals. Goal refinement stops when the responsibility for satisfying goals can be assigned to an agent. Such leave goals define functional requirements for the system. In the example, the IoT infrastructure is responsible for achieving the reliability requirement (keep the packet loss below a required threshold); the operator is responsible for achieving the efficiency requirement (optimize the energy consumed by the motes).

The obstacles in the goal models of Figure 7.1 represent uncertainties that affect the realization of goals. Similar to goals, obstacles can be combined and their effects propagate uncertainty up in the graph toward the top-level goal. Lower-level uncertainties represent concrete sources of uncertainty, while higher-level uncertainties represent the impact of the sources of uncertainty on higher-level goals.

Assessing uncertainties in a goal model may indicate that the system goals cannot be achieved under all conditions. This calls for additional mechanisms to mitigate uncertainties. The goal model on the right illustrates two such mechanisms.

The first mechanism is adding a sub-goal to mitigate an uncertainty, marked with (1) in Figure 7.1. This mechanism requires adding self-adaptive behavior to the system that is responsible for achieving the sub-goal. In the example, a sub-goal *Achieve[SNR ≤ 0]* is added as a sub-goal of *Maintain[Packet Loss Threshold]*, which mitigates the network interference that affects the goal. The new sub-goal refines the requirement *Achieve[Transmission Power Settings]* by requiring power settings for the transmission of packets by the motes that keep the *SNR* for each link below 0. The responsibility for achieving this requirement is allocated to the IoT managing system.

If the refinement of a sub-goal is not sufficient to mitigate the uncertainty and a partial satisfaction of the goal is acceptable, then the goal may be relaxed in order to mitigate the uncertainty, marked with (2) in the example of Figure 7.1. The *Maintain[Packet Loss Threshold]* goal is therefore relaxed to *Maintain[Average Packet Loss Threshold IN 12 hours]*, which allows the self-adaptive system temporarily not to comply with the original requirement in order to handle the uncertainty caused by network interference.

In addition to these uncertainty mitigation mechanisms, management tasks assigned to humans can also be delegated to the managing system. For instance, the responsibility for the requirement *Achieve[Efficiency]* can be allocated to the IoT managing system, releasing the operator from this assignment (this delegation is not shown in Figure 7.1). Instead of letting an operator optimize the power settings of links in the network based on trial-and-error, the IoT managing system will become responsible for optimizing the power settings of the motes, avoiding unnecessary energy consumption.

7.1.2.3   A Note on the Realization of Requirements Reflection

Requirements reflection should not be confused with the representation of adaptation goals as first-class citizens in self-adaptive systems. In the conceptual model of self-adaptive systems (see Figure 1.2), the adaptation of a managed system is driven by a set of adaptation goals. Adaptation goals are explicit runtime entities that represent stakeholder requirements. The designer of a self-adaptive system is responsible for translating the stakeholder requirements into adaptation goals in a machine-readable format that can be used by the feedback loop of the managing system to make adaptation decisions.

With requirements reflection, the *requirements* of a self-adaptive system become first-class citizens, which goes beyond the mere representation of adaptation goals as traditionally used by a managing system (for instance in the form of a set of rules or a utility function). Since requirements are expressed by stakeholders, requirements reflection paves the way for engaging stakeholders in self-adaptive behavior. In particular, requirements reflection provides a basis for communicating the rationale for adaptation decisions a self-adaptive system makes to stakeholders – or more strongly, it allows stakeholders to be actively involved in the decision-making process of a self-adaptive system. Engaging stakeholders in self-adaptive behavior can contribute to the trustworthiness of a self-adaptive system. However, realizing full support for requirements reflection is a challenging undertaking. The goal modeling approach explained in this section offers a basis for a model-driven engineering approach for self-adaptive systems that can support requirements reflection. Such an approach can start from a conceptual model that captures the uncertainties of a self-adaptive system and systematically progresses from requirements and goals to their designs, implementation, and deployment. Requirements reflection can then be realized by keeping the goal model alive throughout this process, into the runtime. Note that this is a necessary but not a sufficient condition to support stakeholder engagement in self-adaptive behavior. Depending on the concrete purpose of stakeholder engagement, this requires at least representation languages, interaction protocols, and supporting infrastructure. For instance, if one wants to involve an operator in the decision of adapting the configuration of DeltaIoT, for instance when a security threat is detected, this will require a notation to communicate the issue to the operator, a protocol that determines the steps that need to be taken to handle the issue, and an infrastructure that supports this interaction.

7.2 Meta-Requirements for Self-Adaptation

A second approach to requirements engineering of self-adaptive systems starts from the question:

> What is the requirements problem a feedback loop for self-adaptation is intended to solve?

In answer to the question, the second approach takes a stance on the requirements problem to be solved by a feedback loop from the point of view of the runtime state/success/failure of the other requirements of the system, i.e. the requirements of the managed system. Hence, the requirements of a managing system are meta-requirements of the requirements of the managed system. These meta-requirements represent the concerns of the managing system in the conceptual model of self-adaptive systems.

This section describes a modeling approach to specifying meta-requirements for feedback loops. This approach is based on the notions of awareness requirements and evolution requirements, which are well known modeling concepts to specify meta-requirements of self-adaptive systems. The two types of requirements complement one another: awareness requirements indicate the situations that require adaptation and evolution requirements prescribe what to do in these situations. By combining both, the engineer can specify the requirements for a feedback loop that operationalizes adaptation at runtime.

7.2.1 Awareness Requirements

Awareness requirements express situations where deviations from the regular system requirements are tolerated by the stakeholders. In particular, an awareness requirement specifies the degree of success of another requirement that is acceptable or the degree of failure that can be tolerated.

Table 7.2 shows different types of awareness requirements.

A regular awareness requirement refers to another requirement that should never fail. An example in DeltaIoT is:

> AR1: *Fail-safe Operation: NeverFail*

This awareness requirement states that the system should never fail to ensure fail-safe operation when needed.

An aggregate awareness requirement imposes constraints on the success/failure rate of another requirement. An example in DeltaIoT is:

> AR2: *Packet Loss ≤ 10%: SuccessRate(100% , 12h)*

This awareness requirement states that the system should keep the packet loss below 10% over periods of 12 hours.

A trend awareness requirement compares the success rates over a number of periods. A delta awareness requirement, on the other hand, specifies acceptable thresholds for the satisfaction of other requirements, such as achievement time. An example for DeltaIoT could be:

> AR3: *ComparableDelta(SetFailsafe, Transmit Packets, 1 cycle)*

This awareness requirement states that when needed, the fail-safe state should be set before packets are transmitted within one communication cycle of the system.

Table 7.2 Types of awareness requirements.

Type	Short description
Regular	A requirement that should never fail.
Aggregate	Imposes a constraint on the success/failure rate of another requirement, which can refer to a frequency, an interval, or a bound on the success/failure.
Trend	Enables comparing success/failure rates over a number of periods in order to specify how success/failure rates of a requirement evolve over time.
Delta	Enables the specification of acceptable thresholds for the fulfillment of a requirement.
Meta	An awareness requirement about another awareness requirement.

Finally, a meta awareness requirement is an awareness requirement on another awareness requirement. Meta awareness requirements can be of any other type of awareness requirement.

Awareness requirements need to be achieved at runtime when the concrete conditions under which they apply can be tracked. Achieving awareness requirements may require the system to adapt. This requires a first-class representation of the requirements, enabling the system to reason about their achievement and take appropriate action when needed. To that end, similar to specification languages for relaxing requirements, the language primitives need to have well defined semantics. The following excerpt shows how example requirement AR3 can be specified in a notation based on the Object Constraint Language (OCL[1]):

```
/** Semantics AR3: ComparableDelta(SetFailsafe, Transmit Packets, 1 cycle) */
1: context TransmitPackets
2:    def: related: Set = SetFailsafe.allInstances() →
3:        select(t | t.arguments('callID') = self.arguments('callID'))
4:    inv AR3: eventually(related → size() == 1) and
5:        always(related → forAll(t: oclInState(Succeeded) and
6:        t.cycle.difference(self.cycle, CYCLES) ≤ 1)
```

The first line states that the context for awareness requirement AR3 is task *TransmitPackets*. The definition (lines 2 and 3) uses the *select()* operator to separate single instances of the related task *SetFailsafe* from the set of instances. The invariant (lines 4 to 6) states that eventually the related set should have exactly one element, which should both be successful and finish its execution within one cycle (recall that packets are transmitted in cycles, using time synchronized communication). Hence, this delta awareness requirement specifies invariants over single instances of the requirements.

Awareness requirements can be graphically represented as illustrated in Figure 7.2. The figure shows an excerpt of a goal model for DeltaIoT with awareness requirements AR1, AR2, and AR3. The specific notation is based on GORE, short for Goal-Oriented Requirements Engineering, a well-known approach for goal modeling. GORE relies on primitives concepts such as: goals, softgoals, and quality constraints. Goals are decomposed until they reach a level of detail that allows a task to be associated with an actor (human or system) that is responsible for performing the task and fulfilling the associated goal. Softgoals are special types of goals that relate to quality properties, which can be expressed as quality constraints. Tasks can contribute to the satisfaction of softgoals.

7.2.2 Evolution Requirements

Evolution requirements prescribe the changes that need to be applied in situations where awareness requirements apply. An evolution requirement can be specified as a sequence of operators. These operators can either change elements or instances of the goal model or they have an effect on the managed system that is subject to adaptation and/or the managing

---

1 ISO/IEC 19507:2012(en): Information Technology – Object Management Group Object Constraint Language (OCL) – https://www.iso.org/obp/ui

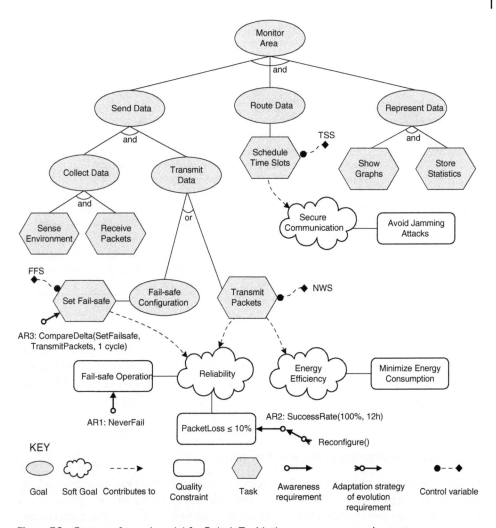

Figure 7.2 Excerpt of a goal model for DeltaIoT with three awareness requirements.

system that adapts the managed system. Table 7.3 illustrates a number example operators that can be used to express evolution requirements.

The sequence of operations of an evolution requirement can be captured in an adaptation strategy. We illustrate an adaptation strategy of an evolution requirement for DeltaIoT:

> //** c: current configuration; NWS: network settings; AG: adaptation goals */
> Reconfigure(algo: bestAdaptationOption(c, NWS, AG), ar:SuccessRate(100, 12)) {
>   c' = find-config(algo, ar);
>   apply-config(c');
> }

The adaptation strategy *Reconfigure* defines a sequence of two operations to deal with failures of the awareness requirement *SuccessRate(100, 12)* (i.e. AR2). The strategy takes

Table 7.3 Evolution requirement operators with effects on managed and managing system.

Operator	Short description of effect
apply-config(C)	The managed system should change from its current configuration to the specified configuration C.
change-param(R, p, v)	The managed system should change the parameter p to the value v for all future executions of requirement R.
find-config(algo, ar)	The managing system should execute algorithm *algo* to find a new configuration to reconfigure the managed system when awareness requirement *ar* fails. The managing system should provide the algorithm all data that is required to find a suitable configuration.
disable(R)	The managed system should stop trying to satisfy requirement R from now on. If R is an awareness requirement, the managing system should stop evaluating it.
enable(R)	The managed system should resume trying to satisfy requirement R from now on. If R is an awareness requirement, the managing system should resume evaluating it.

as arguments an algorithm *algo* to find a new configuration and the awareness requirement *ar* that triggered the strategy. The algorithm uses the current configuration c, the network settings *NWS* (i.e. the control variables) that define the adaptation options to select a new configuration, and the adaptation goals *AG* to determine the best option. The strategy, which is executed when the awareness requirement fails, will find a new configuration that satisfies the goals and adapt the managed system accordingly.

7.2.3 Operationalization of Meta-requirements

Similarly to relaxed requirements, the operationalization of meta-requirements and the realization of their effects require the system to be enhanced with self-adaptive behavior. For meta-requirements realized through awareness requirements and evolution requirements, the central artifacts that are needed for their operationalization are: a runtime representation of the goal model of the system, a monitoring framework to track the state of the awareness requirements, mitigating mechanisms that implement the adaptation strategies of the evolution requirements for each awareness requirement, and system support to enact the operations of the adaptation strategies.

Different strategies can be applied to implement these aspects. The goal model can be implemented using object-oriented techniques, where the element types of the goal model are implemented as classes. Each concrete element can then be instantiated as an object at runtime. To monitor the state of the awareness requirements, the system and its context need to be equipped with probes that track the different parameters of the awareness requirements; for instance, the value of the packet loss over time for AR2. Evaluation requirements can, for instance, be implemented as event-condition-action rules. The events that trigger the rules are the state changes of the awareness requirements; for example, a change in the observed packet loss. The conditions that trigger the rules define the constraints of the awareness requirement; for instance, the observed packet loss exceeds the allowed threshold, either in one cycle or accumulated over a particular time window. The actions of the event-condition-actions rule consist of the sequence of operations that is defined by the adaptation strategy per awareness requirement. The operations result in adaptations of the managed system or changes to the goal model (for example, a goal is temporary disabled). Finally, the system needs to provide effectors that enable enacting the operations. Zanshin is a well-known framework that implements this strategy for the operationalization of meta-requirements.

7.3 Functional Requirements of Feedback Loops

The first two approaches for requirements engineering of self-adaptive systems look at the requirements problem from the viewpoint of the purpose of self-adaptation as expressed by the stakeholders. These approaches provide means to express and operationalize the stakeholder requirements that need to be solved by the self-adaptive behavior that is added to the system. These approaches are centered around making requirements less rigid in order to deal with uncertainties.

The third approach provides a complementary viewpoint and focuses on requirements about the functional behavior of feedback loops. In particular, this approach looks at requirements that need to be satisfied to guarantee that the MAPE components of a feedback loop realize their functions correctly. An intuitive example of such requirement is: the analysis component should correctly identify errors based on the monitored data. Another one is: when the planner component has composed a plan to adapt the managed system, the executor component should eventually execute this plan. Achieving such requirements is important to guarantee correct adaptation capabilities of a feedback loop.

A common approach to ensure correctness of the behavior of a software system is testing its implementation. Testing runs the software against a set of test cases with the intent of finding errors or other defects in the software. These errors or defects can then be corrected in order to make the software fit for use. Software testing can determine the correctness of software based on the set of test cases (oracle) that are used. Testing contrasts with formal verification, which aims to guarantee that the software that is built fully satisfies all the expected requirements. Formal verification employs mathematical techniques to demonstrate that a system is correct with respect to a set of properties. It is important to note that formal verification is not applied to the system itself, but to a formal model of the system. Hence, verification needs to be complemented by validation to confirm that the software has been built according to what the stakeholders really require.

The third approach for requirements engineering of self-adaptive systems combines design-time correct-by-construction modeling of the feedback loop with direct deployment and execution of the verified feedback loop model to realize self-adaptation. Direct execution of the verified model of the feedback loop during operation preserves the guarantees obtained during design and avoids model-to-code translation. The approach is based on ActivFORMS, a well-known approach that provides guarantees for the functional requirements of MAPE-based feedback loops.

7.3.1 Design and Verify Feedback Loop Model

Correct-by-construction modeling of the feedback loop requires: (i) formal models of the MAPE-K components of the feedback loop; (ii) formally specified properties of the functional requirements of the feedback loop, and (iii) a model verifier to perform the actual verification and check that the model complies with the properties.

Figure 7.3 shows example models of the monitor and analyzer components for DeltaIoT.

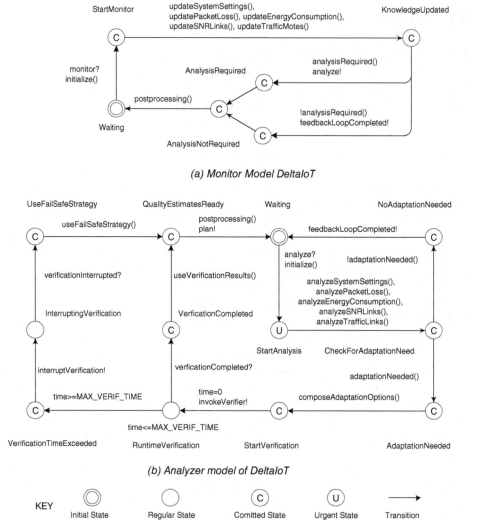

(a) Monitor Model DeltaIoT

(b) Analyzer model of DeltaIoT

Figure 7.3 Examples of feedback loop models for DeltaIoT: Monitor (top) and Analyzer (bottom).

The models are specified as timed automata; Section 6.4.4 briefly explains the basics of this formalism. The two models are aligned with the work flow models of the primary functions of self-adaptation, see Section 4.4 – more specifically Figure 4.9 for the monitor model and Figure 4.10 for the analyzer model.

In short, when triggered by the probe, the monitor updates the different elements of the knowledge, including the actual values of the quality properties – i.e. the packet loss and energy consumption – and the values of the uncertainties – i.e. the signal-to-noise ratio and the traffic generated by the motes. The monitor then checks whether the new values require further analysis. If not, the feedback loop completes; otherwise, the analyzer is triggered, which performs an in-depth analysis of the different properties to decide whether adaptation is required or not. If adaptation is required, the adaptation options are composed, and each option is verified by the external verifier to determine its expected values for packet loss and energy consumption. The analysis results are then used by the planner and executor model to select the best option and adapt the managed system accordingly. The analyzer defines a maximum time period to perform the verification. If this period is exceeded, the verification is interrupted and a fail-safe strategy is applied.

Timed automata models allow the specification of properties over MAPE models. These properties represent functional requirements that the MAPE models should comply with to guarantee the correctness of the feedback loop behavior. Such properties can be specified in computational tree logic (CTL), describing state and path formulae that can be verified, such as reachability (a system should/can/cannot/… reach particular states), liveness (something eventually will hold), etc. The following examples show three such properties for the monitor and analyzer models:

> //** Example properties specified in CTL; a right arrow indicates "eventually" */
> P1. *Monitor.AnalysisRequired → Analyzer.CheckForAdaptationNeed*
> P2. *Analyzer.AdaptationNeeded → Verifier.VerificationCompleted*
> P3. *Analyzer.RuntimeVerification && Analyzer.time >=*
>   *Analyzer.MAX_VERIF_TIME → Analyzer.UseFailSafeStrategy*

Properties P1 and P2 have obvious semantics. Property P3 states that the analyzer should use the fail-safe strategy if the maximum verification time is reached. *Analyzer.time* refers to the actual value of the clock variable *time*, and *Analyzer.MAX_VERIF_TIME* refers to a constant the defines the maximum allowed verification time.

While these properties may seem trivial when inspecting the MAPE models, it is important to note that the verification of these properties allows checking the correctness of the specification of all the model elements, including functions, guards, and invariants.

It is important to note that the verification of the behavior of the MAPE models requires stub models for elements that are external to the verified models. This includes stubs for the probes and effectors, and the runtime verifier. These stubs need to capture the essential behaviors of these external elements and produce all the input that is required to ensure that all the relevant paths through the MAPE models are exercised in order to verify the different properties. For instance, to verify property P3, the models should use input that will generate the condition that the maximum verification time will be reached.

7.3.2 Deploy and Execute Verified Feedback Loop Model

Once the models of the feedback loop are verified, they are ready to be deployed to realize self-adaptation. Direct execution of the feedback loop models requires a model execution engine. If this engine executes the feedback loop model correctly, i.e. according to the semantics of the modeling language, it ensures that the guarantees for the correct behavior of the feedback loop model obtained during design and verification are preserved. The deployed feedback loop models need to be connected with probes and effectors as well as with the verifier that is used to analyze the adaptation options during operation. To preserve the verification results obtained during the verification of the MAPE models, it is essential that these connectors are implemented correctly. Depending on the demands of the problem at hand, the correctness of the model execution engine and the connectors for the MAPE models can be obtained using various techniques, ranging from extensive testing to formal proof.

Figure 7.4 shows an overview of a runtime architecture that realizes the third approach for requirements engineering of self-adaptive systems.

The architecture aligns with the two lowest layers of the three-layer reference model for self-adaptive systems, see Section 5.2. The managed system that is subject to adaptation is instrumented with probes and effectors. The managing system comprises the feedback loop, which is realized by means of a set of formally verified MAPE models that interact with the knowledge models. The analyzer is supported by a model verifier to analyze the adaptation

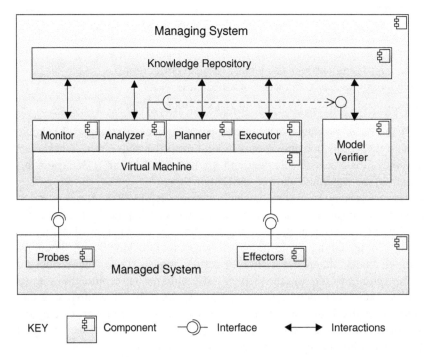

Figure 7.4 Overview of a runtime architecture that realizes the third approach for requirements engineering of self-adaptive systems.

options. The MAPE models are directly executed by the model execution engine, realizing the adaptation of the managed system.

7.4 Summary

Understanding requirements and handling them as first-class citizens is essential for the systematic engineering of self-adaptive systems. Compared to requirements engineering of traditional systems, in self-adaptive systems requirements need to address uncertainty.

Requirements of self-adaptive systems can be specified using specification languages that are extended with operators that allow handling uncertainty. Such operators support the relaxation of requirements in time (temporal) or in terms of quantities (ordinal).

Goal models offer an approach to mitigating uncertainties. Two mechanisms to do so are relaxing an existing goal and adding a sub-goal that is achieved by the adaptive behavior. In addition, management tasks assigned to humans can also be delegated to a managing system.

Requirements reflection makes requirements first-class citizens in self-adaptive systems. One approach toward a basic realization of requirements reflection is applying model-driven engineering centered around goal models. However, realizing the full potential of requirements reflection, in particular engaging stakeholders in the decision making of self-adaptive behavior, is a challenging endeavor.

Meta-requirements deal with the requirements problem a feedback loop is intended to solve. Meta-requirements relate to the state/success/failure of the requirements of the managed system.

Awareness requirements and evolution requirements offer a combined solution to realize meta-requirements. Awareness requirements indicate situations that require adaptation, while evolution requirements specify what to do in such situations.

Different types of awareness requirements are: regular, aggregate, trend, delta, and meta. Awareness requirements can be operationalized through a first-class goal model. To keep the model up-to-date, the relevant elements and parameters of the managed system and its environment need to be tracked at runtime through sensors.

An evolution requirement can be specified as a sequence of operators that work either on the goal model of the system or on the managed system. An adaptation strategy can be defined as a sequence of operators. Evolution requirements can be implemented using event-condition-action rules.

Functional requirements of feedback loops refer to the behavior of a feedback loop. These requirements need to be achieved to ensure that the MAPE models realize their functions correctly. Testing is a common approach to checking functional requirements, but the guarantees obtained through testing are confined to the oracle used. Formal verification provides the means to demonstrate the correctness of a managing system with respect to a set of properties. However, formal verification is bound to the models used.

Combining correct-by-construction modeling of a feedback loop with deployment and direct execution of the models to realize self-adaptation allows the guarantees for the correct behavior of a feedback loop to be preserved. However, this approach requires a trusted virtual machine to execute the models during operation.

Table 7.4 summarizes the key insights of Wave IV.

Table 7.4 Key insights of Wave IV: Requirements-Driven Adaptation.

• Requirements-driven adaptation takes a stakeholder perspective on the requirements problem of self-adaptation.

• The requirements problem of a self-adaptive system is twofold: (i) requirements about the purpose of self-adaptive behavior and (ii) requirements about the correct behavior of the feedback loop.

• Due to uncertainties, a self-adaptive system may not be able to achieve its requirements at all times.

• One approach to deal with uncertainties is tolerating a relaxation in the non-critical requirements of a self-adaptive system. Specification languages with relaxation operators offer a solution to this.

• Meta-requirements are another approach to deal with uncertainty in a self-adaptive system. Awareness and evolution requirements offer a complementary solution to meta-requirements.

• Combining correct-by-construction modeling of a feedback loop with direct execution of the models to realize self-adaptation is one approach to guarantee correct behavior of a feedback loop.

7.5 Exercises

7.1 Goal model for robotic system: level D

Setting. Consider a robotic system in which robots have to perform transportation tasks (pick a load, move, and drop a load) in a warehouse environment with a graph-based routing layout. To move through the warehouse, each robot maintains a representation of the routing layout. The system provides a location service to the robots to track their current positions on the map. The resources in this domain are elements of the map, such as a lane and an edge. We consider two types of uncertainties. First, due to maintenance tasks in the warehouse environment, a lane may need to be temporarily closed. Second, under peak load, for instance when multiple trucks with loads arrive at the same time, additional drop locations may be introduced temporarily to store loads. To that end, operators can manipulate the layout during operation using a set of actions to add and remove a location, an edge, and a destination. Robots can monitor the instructions of the operators and disable or enable particular elements of the layout on the map they use to drive or perform tasks. It is important to note that a requested change of the layout may need to be postponed temporarily until all the required conditions for the robots to change the layout are satisfied. For instance, when a robot is driving to a destination that needs to be removed, this destination can only be removed after the robot has finished its current task. Similarly, a lane can only be closed when no robot is driving on the lane or needs the lane to perform its current task. Hence, the change of the layout in the warehouse will only be effected after the robots that are affected by the change have confirmed the request for the change.

Task. Specify a goal model for the robotic case that includes obstacles representing the uncertainties. Start with a baseline model in which a robot confirms layout changes only when it has completed its current task. Then add mitigation mechanisms that minimize the time a robot needs to confirm layout changes. Explain your solution and define the responsibilities of any type of self-adaptation agent that is used in your solution.

Additional material. Further information about the robotic case can be found in [104].

7.2 Enhanced goal model for DeltaIoT: level D

Consider the goal model for DeltaIoT with the awareness requirements shown in Figure 7.2. Extend this model with an additional awareness requirement that ensures that the average latency of packets over periods of 12 hours is less than 5% of the cycle time of the IoT network. Specify the semantics of the new awareness requirement in OCL. Then define an adaptation strategy that prescribes how the changes need to be applied in situations where the awareness requirement applies. Assess your solution. The same latency requirement was applied in [202]. Further information about awareness requirements and adaptation strategies can be found in [184].

7.3 Goal model for "Feed me, Feed me" artifact: level D

Setting. "Feed me, Feed me" is an artifact of an IoT-based ecosystem that supports food security at different levels of granularity: individuals, families, cities, and nations. The artifact comes with two animated videos that describe the requirements of the system using two comparable scenarios that highlight the positive and negative aspects of the same situation. We focus here at the level of cities, which is centered on the interactions between users and supermarkets, with an emphasis on the practical use or misuse of the real-time data collected by the IoT system. In particular, users share information about grocery requirements through a platform. Supermarkets can gather this real time data to better manage their stock and inventory, and provide users with targeted offers and promotions. However, the data also allows supermarkets to adapt prices depending on the demand, which raises concerns for users about the privacy of the data they share.

Task. Watch the videos of the artifact, and pay special attention to the level of cities where users share information with supermarkets. Identify the goals of the system, and specify an initial goal model. Analyze the tradeoffs for users between the exposure of personal data via the platform (privacy concerns) and the benefits offered by the supermarkets. Translate these tradeoffs to obstacles that may affect the goals, and incorporate them in the goal model. Identify how goals can be relaxed and/or subgoals can be added to mitigate the obstacles. Repeat the task but enhance the initial goal model now with awareness requirements and adaptation strategies to deal with the privacy concerns of users. Explain both solutions and compare them. Appraise both solutions and draw conclusions.

Additional material. For the "Feed me, Feed me" artifact, see [26]. For the movies and additional information of the artifact, see [27] and [25].

7.4 Reliable heating control in a smart home: level D

Setting. Consider a simple smart home system that controls the heating in a house. The system periodically senses the temperature inside the house using four sensors. A controller processes this data to adjust the heating regulator as needed. However, sensors may not work properly and may produce incorrect data or no data at all. The owner of the house wants to make sure that the controller only considers the data of sensors that are working correctly. To that end, a feedback loop is added to the system that observes whether the sensors are working correctly. A sensor is considered to be working incorrectly if it produces no data or if the value of the temperature it produces diverges too much from the values produced by the other sensors. When the feedback loop detects that a sensor is not working correctly, it adapts the setting of the controller such that the data of that sensor is not included when determining the control actions of the heating system. As soon as the sensor is producing correct data again, the feedback loop adapts the setting of the controller such that the data of the sensor is included again.

Task. Specify a model of the simple smart home system with a basic controller using timed automata. Add a probe automaton to the control system that allows the sensor values to be observed. Add an effector automaton that allows data of individual sensors to be excluded or included. Now design a MAPE-K feedback loop that deals with sensors that are working incorrectly. Generate a sequence of measurements of the sensors, and include some sensor failures. Now specify a set of correctness properties for the feedback loop, and verify these properties using the Uppaal verifier. Then test your solution using the Uppaal simulator. Assess your solution.

Additional material. The Uppaal tool with all the necessary documentation can be downloaded via the Uppaal website [189].

7.6 Bibliographic Notes

The pioneering book of R. Thayer and M. Dorfman covered the five phases of software requirements engineering: elicitation, analysis, specification, verification, and management [187]. Seminal papers on requirements engineering were presented at the Future of Software Engineering Symposium in 2000 [148] and in 2007 [49]. I. Jureta et al. elaborate on the fundamental requirements problem and define a core ontology [111].

J. Whittle et al. introduced the RELAX language, which introduced a set of operators to tolerate a relaxation of non-critical requirements [214]. Background on fuzzy temporal logic, which is used to define the semantics of RELAX operators, can be found in [141]. The basic work on goal-based modeling with KAOS is described in [124]. B Cheng et al. demonstrated how the RELAX specification language can be used to specify more flexible requirements within a goal model to handle uncertainty [51]. FLAGS, short for Fuzzy Live Adaptive Goals for Self-adaptive systems, is another approach for requirements relaxation that generalizes a KAOS model by adding fuzzy-based adaptive goals to realize self-adaptation [16].

D. Berry and B. Cheng argued for four levels of requirements engineering in a dynamic adaptive system [30].

N. Bencomo et al. elaborated on requirements reflection, which considers requirements as first-class runtime entities [22].

V. Silva Souza et al. introduced awareness requirements for self-adaptive systems [183]. In follow up work, the authors complemented awareness requirements with evolution requirements [184]. For more information about the Zanshin framework, see [185].

ActivFORMS was introduced in [101]. S. Shevtsov et al. compared ActivFORMS with a control-based approach to self-adaptation for the TAS exemplar [177]. For a more elaborated description of ActivFORMS, see [202].

8

Wave V: Guarantees Under Uncertainties

The fifth wave puts the emphasis on "taming" uncertainty, i.e. providing *guarantees* for the compliance of a self-adaptive system to its adaptation goals despite the uncertainty the system is exposed to. The third wave, Runtime Models, which aimed at managing the complexity of concrete designs of self-adaptive systems, brought the topic of uncertainty in the picture through the use of probabilistic models that capture uncertainties. These models can be used at runtime to reason about the running system and adapt it as needed to achieve its goals. The fourth wave, Requirements-Driven Adaptation, which focused on the requirements problem of self-adaptive systems, fueled the importance of uncertainty in self-adaptive systems. The insights obtained from these two waves were the drivers for the fifth wave, which introduces a shift in the motivation for self-adaptation: uncertainty becomes a central driver for self-adaptation.

Uncertainty in a self-adaptive system might originate from disparate sources, including the environment in which the system operates, the behavior of humans involved in the system, the goals of the system, the system itself that is managed, but also the feedback loop and the runtime models the system uses to realize the self-adaptive behavior. For self-adaptive systems with strict requirements, this leads to a dichotomy between the uncertainties on the one hand and guarantees for the system goals on the other hand. "Taming uncertainty" refers to the need to deal with this dichotomy. Hence, the focus of the fifth wave is on achieving compliance of a self-adaptive system with the adaptation goals that it needs to achieve, despite of the uncertainties the running system is subject to.

In general, guarantees for requirements compliance of a software system are usually provided offline, i.e. before the system is put in operation. Depending on the criticality of the system, the evidence that needs to be provided for the guarantees can be obtained by a variety of techniques, ranging from code inspection and testing to model checking and formal proof.

Due to uncertainties, it is often hard or even impossible to obtain the evidence for requirements compliance of a self-adaptive system through offline techniques only. Hence, self-adaptive systems require an enduring process that continuously provides new evidence to deal with the uncertainties that the system faces over its lifetime, from inception to and throughout operation in the real world. This process introduces a blur in the role of the traditional engineering phases of software systems, where evidence of

An Introduction to Self-Adaptive Systems: A Contemporary Software Engineering Perspective,
First Edition. Danny Weyns.
© 2021 John Wiley & Sons Ltd. Published 2021 by John Wiley & Sons Ltd.

goal compliance of the system at runtime is obtained during offline development and evolution activities. Self-adaptive systems require a seamless integration of evidence obtained through human-driven offline activities with machine-driven online activities.

We start this chapter by elaborating on the notion of uncertainty in self-adaptive systems and outline different sources of uncertainty. This sets the scene for the study of a characteristic approach to taming uncertainties in self-adaptive systems. This approach, which relies on formal verification of adaptation options at runtime, follows a two-step process. In the first step, the adaptation options are analyzed, and in the second step, the best adaptation option is selected to adapt the managed system. We present three concrete instances of this approach.

The first instance applies quantitative verification at runtime to analyze the adaptation options. Quantitative verification is a set of mathematically-based techniques that can be used for analyzing quality properties of systems that exhibit stochastic behavior. Through exhaustive verification, this approach provides strong evidence for requirements compliance of the adaptation options that are selected to adjust or reconfigure the self-adaptive system. In this book, we use the term quantitative verification interchangeably with probabilistic model checking.

A second concrete instance applies statistical model checking at runtime to analyze the adaptation options. This approach combines simulation at runtime with statistical techniques. Compared to exhaustive techniques, this approach is more efficient in terms of the resources and time it requires. However, the evidence for requirements compliance of a self-adaptive system provided by the second approach is bounded by a certain level of accuracy and confidence.

The third concrete instance applies probabilistic model checking at runtime to make proactive adaptation decisions. Instead of reacting to changes in the short term, which may result in sub-optimal adaptation decisions, this approach makes adaptation decisions with a look-ahead horizon. The third approach integrates the analysis of adaptation options with the selection of the best option for adapting the managed system.

The chapter concludes with an overview of an integrated engineering process to tame uncertainty in self-adaptive systems. This perpetual process seamlessly integrates evidence that is obtained through offline activities with new evidence that is obtained during operation. The integrated process is based on ENTRUST, short for Engineering of Trustworthy Self-adaptive Software, a pioneering approach to engineering trustworthy self-adaptive software systems.

LEARNING OUTCOMES

- To explain the notion of uncertainty.
- To discuss and illustrate different classes of uncertainty sources.
- To motivate and illustrate the need for guarantees under uncertainty.
- To explain and illustrate the principles of exhaustive verification to tame uncertainty.
- To explain and illustrate the principles of statistical verification to tame uncertainty.
- To explain and illustrate proactive decision-making using probabilistic model checking.

- To explain how verification and validation in self-adaptation complement one another.
- To apply different stages of the integrated process to tame uncertainty on a concrete case.

8.1 Uncertainties in Self-Adaptive Systems

Uncertainty in computing systems has been described in various ways, but a common dominator in these descriptions is related to the knowledge that is available to perform the tasks at hand. Uncertainty may refer to the absence of knowledge, the inadequacy of knowledge, and the difference between the knowledge required to perform a task and the knowledge already possessed, among other elements. In general, a distinction can be made between aleatoric uncertainty, which refers to imprecision of knowledge (the main focus in self-adaptation research), and epistemic uncertainty, which refers to the lack of knowledge. We use the following working definition:

> Uncertainty in self-adaptive systems is any deviation of deterministic knowledge that may reduce the confidence of adaptation decisions made based on the knowledge.

Without proper mitigation of uncertainty, adaptation decisions may be inaccurate, unreliable, or inconclusive. This in turn will affect the operation of the self-adaptive system and result in degrading quality or even worse – for instance the violation of safety properties of a system. Hence, recognizing the presence of uncertainty and managing it can mitigate its potentially negative effects and increase the level of trust in a given self-adaptive system. Ignoring uncertainties may lead to unsupported claims on the system's validity or generalization of claims beyond their actual scope.

Table 8.1 classifies the sources of uncertainty into four groups. Per group, a non-exhaustive set of characteristic sources of uncertainty is listed. The sources of uncertainty can manifest themselves at design time, at runtime, or both.

The first group is about uncertainties related to the system itself, which can refer to both the managed system and the managing system. Simplifying assumptions refers to abstractions applied in modeling, which may exclude details that are actually significant. Model drift refers to uncertainty introduced by an emerging discrepancy between changes in the system that is adapted and the corresponding models. Incompleteness manifests when some parts of the system or its model are not fully known, either at design time or at runtime. The lack of knowledge may be gradually addressed at runtime when more information becomes available. For instance, new instances of services that can be used by a Web service application may be discovered at runtime. Future parameter values refers to lack of complete knowledge about the conditions in which the system will operate in the future, for instance the effect of adaptation actions on system parameters. The adaptation functions themselves may be subject to uncertainty, for instance due to imprecision in monitoring, the analysis techniques used, or side effects that adaptation actions may have on the system. Decentralized decision-making may introduce side effects on the global

Table 8.1 Sources of uncertainty.

Group	Source of uncertainty	Brief explanation
System	Simplifying assumptions	Modeling abstractions that introduce some degree of uncertainty.
	Model drift	Misalignment between elements of the system and their representations.
	Incompleteness	Some parts of the system or its model are missing and may be added at runtime.
	Future parameter values	Lack of knowledge about future values of parameters that are relevant for decision-making.
	Adaptation functions	Imperfect monitoring, decision-making, and executing functions for realizing adaptation.
	Decentralization	Lack of accurate knowledge about the system-level effects of local decision making.
	Automatic learning	Learning with imperfect and limited data, or randomness in the model and analysis.
Goals	Requirements elicitation	Elicitation of requirements is known to be problematic in practice.
	Specification of goals	Difficulty of accurately specifying the preferences of stakeholders.
	Future goal changes	Changes in goals due to new customer needs, new regulations, or new market rules.
Context	Execution context	Context model based on monitoring might not accurately determine the context and its evolution.
	Noise in sensing	Sensors/probes are not ideal devices, and they can provide (slightly) inaccurate data.
	Different information sources	Inaccuracy due to composing and integrating data originating from different sources.
Humans	Human in the loop	Human behavior is intrinsically uncertain; it can diverge from the expected behavior.
	Multiple ownership	Parts of the system provided by different stakeholders may be partly unknown.

behavior of the system that are difficult to predict locally. A typical example is unwanted emergent behavior such as oscillations that result from the interaction between different local adaptations. Finally, automatic learning inherently introduces uncertainties as it relies on statistical techniques that work on limited data sets.

The second group is about uncertainties related to the goals of a self-adaptive system. Requirements elicitation concerns defining the system scope, understanding the concrete needs and preferences of stakeholders, and dealing with the volatile nature of requirements, which is particularly relevant for self-adaptation. Accurately specifying the goals of stakeholders is difficult and prone to uncertainty, in particular specifying the tradeoffs between conflicting qualities, which may not be completely known upfront or be affected by the context in which the system operates. Future goal changes refers to changes in the requirements of the system due to new needs of stakeholders, changes in technology, new regulations that come in place, etc.

The third group is about uncertainties in the context of a self-adaptive system. The execution context in which a self-adaptive system operates needs to be represented in a model allowing the managing system to reason about changes. However, such a model may not accurately capture the actual context, for instance due to the inability of sensors to unambiguously determine the context and its evolution over time. Noise in sensing is caused by sensors and monitoring mechanisms that provide slightly different data in successive measures, while the actual value of the monitored data did not change. Different sources of information may introduce uncertainty as the composition and integration of different types of data may introduce inaccuracies.

Finally, the fourth group is about uncertainties related to humans involved in the functioning or operation of a self-adaptive system. Humans may be actively involved in the adaptation work flow, for instance to provide input or support decision-making. However, the behavior of humans is subject to intrinsic uncertainties. Multiple ownership of self-adaptive systems may cause uncertainties as owners may hide certain information of the system, for instance for confidentiality reasons. However, lack of this information may affect the decision-making of adaptations.

In summary, self-adaptive systems can be exposed to a wide range of potential sources of uncertainties. For self-adaptive systems with strict goals, this introduces a paradoxical challenge: how can one provide guarantees for the goals of a system that is exposed to continuous uncertainty? The different approaches that we present now aim to provide an answer to this question.

8.2 Taming Uncertainty with Formal Techniques

Taming uncertainty is centered on the analysis and planning stages of the self-adaptation work flow. A prominent approach to taming uncertainty in self-adaptive systems relies on applying formal techniques at runtime to make adaptation decisions.

This decision-making process consists of two parts. In the first part, the analyzer uses formal verification techniques to evaluate the adaptation options. The focus of this part is in particular on the analysis of adaptation options, see the basic work flow of the analyzer function in Figure 4.10. In the second part, the planner selects the best option based on the verification results and the adaptation goals. The focus of the second part is in particular on determining the best adaptation option, see the basic work flow of the planner function in Figure 4.11.

8.2.1 Analysis of Adaptation Options

Evaluating the adaptation options is the task of the analyzer, which works in tandem with a model verifier to perform the formal verification. Figure 8.1 shows the work flow of formal analysis of the adaptation options.

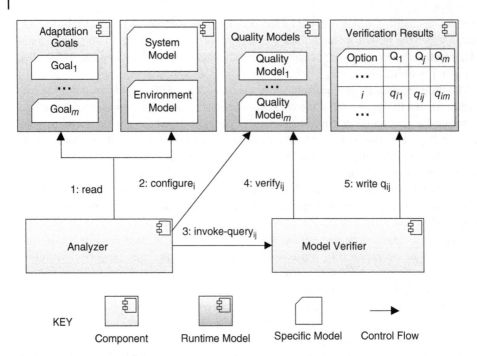

Figure 8.1 Work flow of formal analysis of adaptation options.

The work flow consist of five main steps. In step *1: read*, the analyzer reads the up-to-date runtime models of the managed system, the environment, and the adaptation goals. The model of the managed system is used to determine the adaptation options, each option *j* representing a possible reconfiguration of the managed system.

In step *2: configure$_i$*, the analyzer configures the quality models for adaptation option *i*. Quality models are parameterized models, with parameters that represent the possible settings of the managed system and parameters that represent uncertainties. Configuration boils down to instantiating the settings of the managed system. The values of the settings are derived from the adaptation options that were determined in the first step. The values of the parameters that represent uncertainties are derived from the environment model and possibly the managed system model. Here we use one quality model for each adaptation goal as shown in Figure 8.1. However, in practice, an integrated approach may be used that combines different quality properties in a single model.

In step *3: invoke-query$_{ij}$*, the analyzer invokes the model verifier to verify quality model *QualityModel$_j$*, which was configured with the settings of adaptation option *j* for a property that corresponds to adaptation goal *Goal$_j$*. The query informs the model verifier which model it needs to verify and what property needs to be verified.

In step *4: verify$_{ij}$*, the model verifier performs the formal verification. The result of the verification is an estimate of a quality property Q_j for an adaptation option *i*, with Q_j referring to adaptation option *Goal$_j$*.

Finally, in step *5: write q$_{ij}$*, the model verifier writes the verification result to the knowledge repository. More specifically, the model verifier writes the predicted value of quality q_{ij} to the verification results.

Steps 2, 3, 4, and 5 are repeated for each adaptation option i and each quality property Q_j that corresponds to adaptation goal $Goal_j$.[1] The result of the first part is a list of adaptation options with verification results for each quality property that is subject of adaptation.

8.2.2 Selection of Best Adaptation Option

Selecting the best adaptation option based on the verification results and the adaptation goals is the task of the planner, in addition to then making a plan to enact the selected option. Figure 8.2 shows the work flow of the selection of the best adaptation option.

The work flow consist of five main steps, two of them specifically focusing on the selection of best adaptation options. In step *1: read*, the planner reads the adaptation goals and the verification results produced in the first part.

In step *2: select*, the planner applies a decision-making mechanism to determine the best adaptation option. A classic example of such a decision-making mechanism is a utility function that assigns a value to each adaptation option based on the relative importance of the values of different quality properties as determined during analysis. The option with the highest value is then selected for adaptation. Adaptation options with values below or above

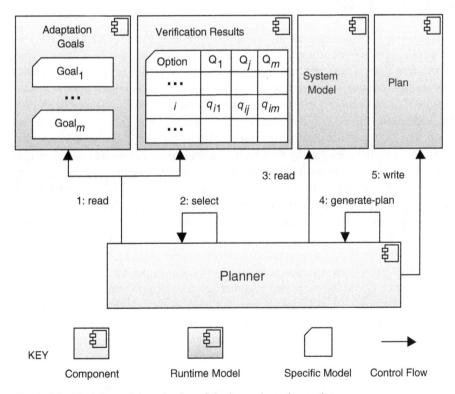

Figure 8.2 Work flow of the selection of the best adaptation option.

---

1 This is the most general case. Different types of optimizations may be possible that do not require verifying all the entries of the matrix. For instance, adaptation options that violate one adaptation goal may be excluded from the verification of future goals.

certain thresholds may be excluded (for instance, an option with a cost that is too high according to the stakeholders). Another simple decision-making mechanism is the use of rules that are applied to the verification results of the adaptation options in a particular order to rank them. The highest-ranking option after the rules are applied is then selected for adaptation.

Once the best adaptation option is selected, the planner will read the current configuration of the system (step *3: read*) to prepare a plan in order to adapt the managed system from its current configuration to the new configuration of the selected adaptation option (step *4: generate-plan*). Finally, this plan is written to the knowledge repository (step *5: write*) where the executor will find it to enact the adaptation actions of the plan and adapt the managed system.

We look now at two concrete instances of the general approach that use formal verification techniques to tame uncertainties in self-adaptive systems.

8.3 Exhaustive Verification to Provide Guarantees for Adaptation Goals

The first concrete approach aims at providing guarantees for the goals of a self-adaptive system using exhaustive verification. A pioneering technique to do so is runtime quantitative verification. Quantitative verification is a well-known technique that is used to verify whether certain quantitative properties hold for a model of a system that exhibits stochastic behavior. The models used in quantitative verification are usually variants of Markov models, annotated with costs and rewards that describe resources and their usage during execution. The properties are commonly expressed in temporal logic extended with probabilistic and reward operators. Examples of such properties are the expected probability that messages get lost during a period of time and the expected time for message delivery. Verification tools are available that exploit clever mechanisms to explore the state transition graph of the system in an efficient way, ensuring complete coverage of the state space.

With runtime quantitative verification, the analyzer of the MAPE feedback loop applies quantitative verification to estimate the qualities of the different adaptation options that are available for adaptation. These verification results are then used to select the best adaptation option that complies with the adaptation goals. The approach we present in this section is based on QoSMOS, short for Quality of Service Management and Optimization of Service-based systems, a well-known approach that relies on quantitative verification at runtime for the analysis of adaptation options.

Figure 8.3 shows the architecture of a MAPE feedback loop that applies exhaustive verification to select adaptation options. The example case is similar to the health assistance system that we used in Chapter 1, see also Figure 1.1.

The service-based system offers users remote access to a composition of Web services. The service work flow is executed by a work flow engine. The functionality of each service in this work flow may be provided by different service providers. The service instances differ in the quality properties they offer, e.g. reliability, response time, cost, etc. Table 8.2 shows an excerpt with characteristic values of the quality properties of concrete service instances.

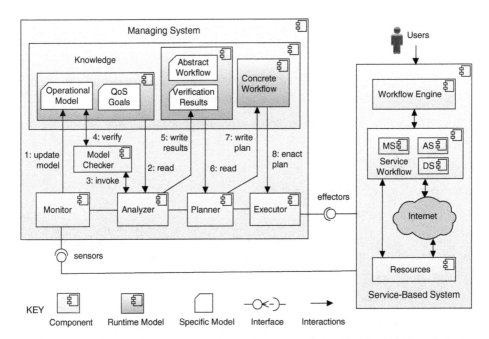

Figure 8.3 Architecture of a feedback loop that uses exhaustive verification for the analysis of adaptation options. Service types: MS: Medical Analysis Service, AS: Alarm Service, and DS: Drug Service.

Table 8.2 Illustration of characteristic values for quality properties of concrete service instances.

Name	Failure rate	Execution time	Cost
MedicalAnalysisService1 (MS1)	0.03	4.1s	9.1c
MedicalAnalysisService2 (MS2)	0.04	3.9s	6.8c
...			
AlarmService1 (AS1)	0.01	2.7s	11.9c
AlarmService2 (AS2)	0.02	0.3s	7.3c
AlarmService3 (AS3)	0.03	0.3s	7.3c
...			
DrugService1 (DS1)	0.02	1.3s	9.8c
DrugService2 (DS2)	0.05	1.7s	6.8c
...			

We consider the following uncertainties:

- System: variations in the availability of concrete services
- Context: variation in the failure rates of concrete services
- Human: variations on usage of services

The adaptation problem is to select concrete services that compose a composite service that offers the highest utility. The utility of the service-based system is defined based on the following requirements:

R1. The probability that a failure of the work flow execution occurs is less than $P_1 = 0.14$.
R2. The cost for executing the work flow should be minimized.

The utility function to select service combinations is defined as follows

$$utility = \sum_{i=1}^{2} w_i \cdot objective_i \tag{8.1}$$

The function $objective_1$ is defined as follows

$$objective_1 = goal(R_1) \tag{8.2}$$

with $goal(R1) : \{false, true\} \rightarrow \{0,1\}$, i.e. $objective_1$ has a value 1 if the quality goal R_1 is satisfied and a value 0 otherwise. The function $objective_2$ is defined as follows

$$objective_2 = -\sum_{i=1}^{3} c_{S_i} \tag{8.3}$$

with c_{S_i} the cost associated with selected service S_i, where S_1 is an instance of the medical analysis service, S_2 an instance of the drug service, and S_3 an instance of the alarm service (see Table 8.2).

The weights of the objectives are defined as follows

$$w_1 = 100 \ and \ w_2 = 1 \tag{8.4}$$

The managing system that interacts with the service-based system through sensors and effectors consists of a classic MAPE loop that uses a set of runtime models to make adaptation decisions.

The *Monitor* component monitors the failure rate of the services (reliability) and the behavior of the service work flow, i.e. the paths that are exercised in the work flow. This information is used to update the *operational model*, which corresponds to the quality model in Figure 8.1. Figure 8.4 shows an excerpt of an operational model that is specified as a Discrete Time Markov Chain (DTMC). In particular, the model illustrates a part the work flow of actions with probabilities assigned to branches in the model. This model can be used to analyze the expected reliability of the service-based system for different combinations of concrete service instances. Failure states are highlighted in gray. The initial estimates of the values of the probabilities are based on input from domain experts. The monitor updates the values at runtime, based on observations of the real behavior. Failure probabilities of service invocations are modeled using variables (a, b, c in the model) because these values depend on the concrete service instances that can be selected by the MAPE loop.

When the *Analyzer* component detects that the quality goals are violated (in the example, the reliability of the service-based system), it will instantiate the parameterized operational model to identify the service configurations that satisfy the quality requirements. To that end, the analyzer uses a *model checker* that takes the DTMC model, instantiates this model for different combinations of service instances, and automatically and exhaustively checks

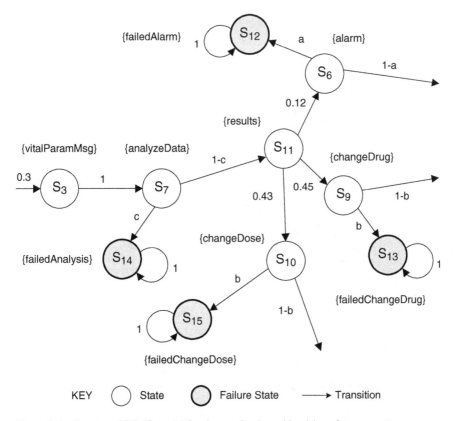

Figure 8.4 Excerpt of DTMC model for the service-based health assistance system.

whether the instantiated model meets the required qualities. To verify the model, the reliability requirement (*R1*) needs to be translated into a formal expression that can be used to perform model checking. For a DTMC model, the reliability requirement can be expressed in Probabilistic Computation Tree Logic (PCTL) as follows

$$R1 : P_{\leq 0.14}[\, \Diamond\text{“}failure\text{”}] \tag{8.5}$$

Informally, this property states that the system shall have a behavior where with a probability of less than 0.14 it is the case that a failure of the health assistance system will occur. A failure of the system occurs when the invocation of any of the concrete services of the work flow fails, i.e. when any of the states *failedAlarm*, *failedAnalysis*, *failedChangeDrug*, or *failedChangeDose* is reached in Figure 8.4. The probability of a failure is determined by the parameters a, b, and c that depend on the concrete services selected by the analyzer.

The property in Equ. (8.5), which refers to an adaptation goal in Figure 8.1, can be used to invoke a query and perform a verification of the operational model.

PRISM is a well-known model checker that supports the analysis of DTMC models for requirements specified in PCTL expressions. The analyzer automatically carries out the analysis of a range of possible configurations of the service-based system (exploring the range of adaptation options) by instantiating the parameters of the operational model for the concrete service instances. Figure 8.5 shows an excerpt of the verification results.

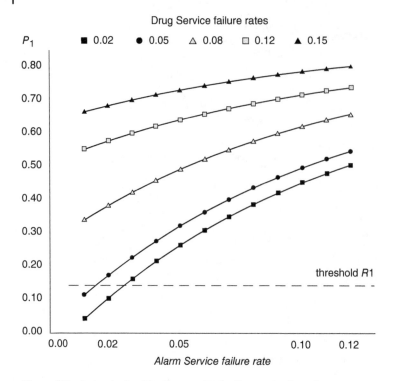

Figure 8.5 Excerpt of verification results for the service-based system.

The results show the probability of failures for a particular set of service combinations. Hence, each combination of service instances per type in Table 8.2 maps to one of the verification results. In this particular case, the medical service instance is fixed and combined with different instances of the alarm service and drug service with failure rates ranging between 0.01 and 0.12 and 0.02 and 0.15, respectively. The service combinations that have a failure rate probability below the threshold of 0.14 comply with requirement R1. All combinations with a failure rate probability above the threshold violate the requirement. In a similar way, the different service combinations can be analyzed for the other requirements.

The *Planner* uses the verification analysis results and the abstract work flow to determine the service combination that will be selected and builds a plan for adapting the work flow of the service-based system accordingly.

To select the combination of services, the planner uses the utility function defined above. This utility function corresponds to the decision-making mechanism that is part of step 2: *select* in Figure 8.2. We illustrate the selection of a new configuration, based on a subset of the adaptation options as shown in Figure 8.6.

For instance, for configuration ①, $objective_1$ is 1 since this configuration complies with requirement R_1 (the probability that the failure rate is below the threshold of 0.14); $objective_2$ is the sum of the costs of $MS1$, $AS1$, and $DS1$ (see Table 8.2): $9.1 + 11.9 + 9.8 = 30.8$; and the utility is $100.1 - 30.8 = 69.2$. For configuration ④ on the other hand, $objective_1$ is 0 since this configuration does not comply with requirement R_1; $objective_2$ is $9.1 + 7.3 + 6.8 = 23.2$;

Figure 8.6 Detail of some of the verification results for the service-based system.

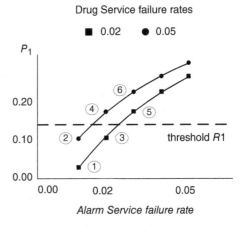

Drug Service failure rates
■ 0.02 ● 0.05

Table 8.3 Determining the utilities of a small subset of configurations shown in Figure 8.5.

Configuration	Selected service instances	$objective_1$	$objective_2$	Utility
①	{MS1, AS1, DS1}	1	30.8	69.2
②	{MS1, AS1, DS2}	1	27.8	72.2
③	{MS1, AS2, DS1}	1	26.2	73.8
④	{MS1, AS2, DS2}	0	23.2	−23.2
⑤	{MS1, AS3, DS1}	0	26.2	−26.2
⑥	{MS1, AS3, DS2}	0	23.2	−23.2

and the utility is $0.0 - 23.2 = -23.2$. If the planner would have to select among these six configurations, it would pick configuration ③ since this has the highest utility.

To enact the new concrete work flow with the service configuration, the planner will create a plan with adaptation actions. Each action maps either to deactivating or enacting a concrete service for the corresponding service type in the work flow. Finally, the *Executor* executes the plan, replacing the old services with the new selected services, which will then be invoked by the work flow engine. This completes an adaptation cycle.

8.4 Statistical Verification to Provide Guarantees for Adaptation Goals

The second concrete approach to provide guarantees for the goals of self-adaptive systems relies on statistical techniques to make adaptation decisions. This approach aims to tackle the inherent state-space explosion problem of exhaustive verification. State-space explosion refers to the exponential growth of the state space when the number of state variables in the system increases, resulting in a dramatic increase of the resources and time that are required to perform verification. The second approach selects adaptation options that realize the adaptation goals in an efficient manner using statistical verification (or statistical model checking) at runtime. Statistical verification performs a series of simulations of a model and then applies statistical techniques on the simulation results to decide whether the configuration satisfies the property with some degree of accuracy and confidence. Statistical verification allows a designer to trade off the required levels of accuracy and confidence of the results with the resources and the time that are used for verification. The approach we present here is based on ActivFORMS, a pioneering approach that applies statistical model checking at runtime.

Figure 8.7 shows the architecture of a MAPE feedback loop that uses statistical model checking to select adaptation options in DeltaIoT.

We consider the default setup of DeltaIoT, with 14 nodes that send sensor data to the gateway via a wireless network. The sensor data is collected by the front end where security staff can monitor the status of buildings and take appropriate action when needed.

The adaptation problem of DeltaIoT is to adapt the transmission power of the motes and the distribution factors of the links such that the adaptation goals are achieved. We consider two types of uncertainties: network interference (context) and fluctuating traffic load (system). The requirements for this scenario are:

R1. The average packet loss of the network should not exceed 10 % over 12 hours.
R2. The energy consumption should be minimized.
R3. The default settings should be applied when no configuration is available to achieve R1.

The managing system forms a feedback loop that interacts with the DeltaIoT system via the management interface.

The *Monitor* component monitors the packet loss of the network, the network interference, and the traffic load generated by the motes. This data is used to update the *Configuration* model. This model comprises a representation of the network with the actual settings (i.e. the transmission power of the motes, distributions of packets to parents), the current values of quality properties (power loss and energy consumption), and the uncertainties (current traffic load of motes and interference of links).

Based on the up-to-date data of the configuration model, the *Analyzer* determines whether an analysis needs to be initiated. Concretely, the analyzer will check the changes of the packet loss and energy consumption of the network, and the SNR of the links and the traffic load generated by the motes. If the changes of any of these variables exceeds a given threshold, the analyzer will initiate an analysis. To that end, the analyzer will use the up-to-date data of the configuration model and the adaptation options to instantiate the quality models.

The *Adaptation Options* for DeltaIoT consist of the set of possible configurations of the network. This set is determined by the network settings of the motes per link (transmission power from 0 to 15 and distribution factor per link from 0 to 100 % in steps of 20 %). The settings of the transmission power for the links are determined such that the SNR for each link is at least zero given the current level of interference along the link; this minimizes packet loss without wasting unnecessary energy. The same power settings are used for all configurations. Combining the possible settings of the distribution factors for all links

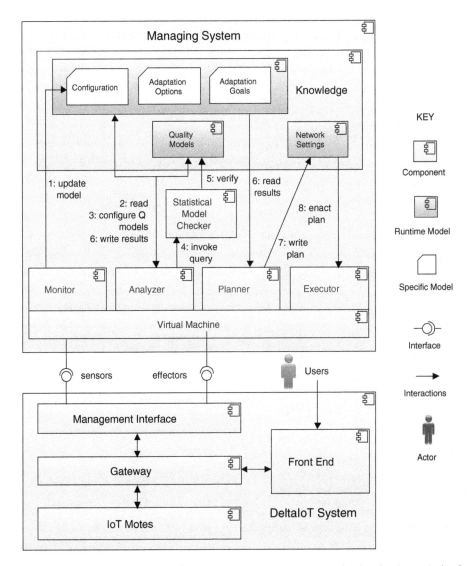

Figure 8.7 Architecture of a feedback loop that uses statistical verification for the analysis of adaptation options in DeltaIoT.

between the 14 motes results in an adaptation space of 216 possible adaptation options in total (including the current configuration).

For each quality property that corresponds to an adaptation goal, the system maintains a *Quality Model.* Figure 8.8 illustrates a runtime model that can be used to predict the energy consumption of different configurations of DeltaIoT.

This quality model consists of three interacting automata: *Mote, Gateway,* and *System.* The automaton that represents the behavior of a mote is instantiated for each mote in the system. This automaton has a number of parameters that can be set dynamically during analysis. In particular, the parameters allow the model to be configured per adaptation

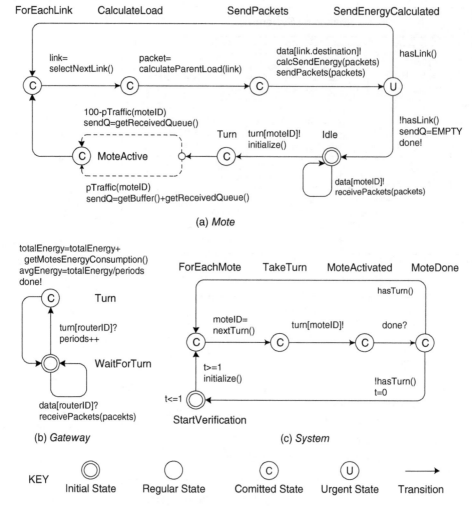

Figure 8.8 Example of a formal model to predict energy consumption of configurations in DeltaIoT.

option by setting the transmission power used by the motes to communicate packets and the distribution factors of the links to send packets to parent motes. Furthermore, the parameters that represent uncertainties can be set according to the up-to-date values of the configuration model. For the quality model in Figure 8.8, this is the actual value of *pTraffic(moteID)*, which represents the traffic that a mote with a given *id* is expected to generate. Once the settings for an adaptation option are set, the estimated energy consumption can be determined.

The analysis of the quality models is performed by the statistical model checker. Uppaal is a well-known model checking suite that supports the verification of timed automata. The following query can be used to estimate energy consumption:

simulate1[<= 30](Gateway.avgEnergy)

The query calculates the estimated average energy consumption for a given configuration over 30 simulation runs. The number of simulation runs will determine the accuracy and confidence of the estimate and can be determined by the designer based on offline experiments of a representative set of model instances.

The model checker starts the verification by activating the System automaton from its initial state. This will activate the motes one by one, *moteID = nextTurn()*, according to the time schedule used. Each mote can then send packets to its parents in the time slots dedicated to it. To that end, the send queue is filled with packets from the buffer that have been generated by the mote itself (depending on the transition taken based on the current value of *pTraffic(moteID)*) and packets in the receive queue that the mote received from its children. Depending on the distribution factors, the load per link is then computed, and the messages are sent to the corresponding parent *(sendPackets(packets))*. When sending packets, the energy required to send the messages is computed *(calcSendEnergy(packets))*. When all motes have had the opportunity to send messages, the gateway receives its turn. The gateway then computes the total energy consumed by the motes both to send and receive messages. This process is repeated a predefined number of times (i.e. 30 times for the query we use here as illustration) to compute an estimated average energy consumption with a required accuracy and confidence.

Figure 8.9 shows an excerpt of the verification results for the adaptation options at a particular point in time.

Each dot on the graph represents an adaptation option with its estimated average values of the two quality properties. The diamond dot represents the configuration of the managed system in use at the time the analysis was performed. This configuration does not comply with the packet loss requirement *R1*; hence, the system should be adapted.

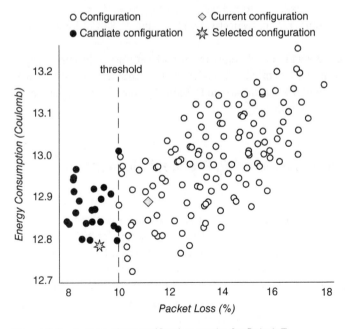

Figure 8.9 Excerpt of the verification results for DeltaIoT.

The *Planner* component uses the analysis results to determine the adaptation option that will be used to adapt the network. To that end, the planner will use the *Adaptation Goals*, which are defined here as boolean functions:

```
int MAX_PACKET_LOSS = 10;
bool satisfactionGoalPacketLoss(Configuration gConf, int MAX_PACKET_LOSS) {
    return gConf.qualities.packetLoss < MAX_PACKET_LOSS;
}
bool optimizationGoalEnergyConsumption(Configuration gConf, Configuration
    tConf) {return tConf.qualities.energyConsumption < gConf.qualities.
    energyConsumption;
}
```

The planner will start by applying the packet loss goal. This goal tests whether the packet loss of a configuration is beneath a given threshold (here defined at 10%). This goal allows the separation of configurations that satisfy the packet loss requirement from those that do not satisfy the requirement. The black dots in Figure 8.9 represent adaptation options that comply with the packet loss goal.

Then the planner will apply the energy consumption goal. This goal tests whether the energy consumption of one configuration is lower as that of another configuration. This goal allows the configuration with the lowest energy consumption to be found. By comparing the adaptation options that comply with the packet loss goal, the adaptation option marked with a star will be selected for adaptation. This option represents the best configuration as it has the minimum predicted energy consumption. The selected adaptation option at this particular point in time is expected to reduce packet loss to 9.5% and energy consumption to 12.75 Coulomb compared to, respectively, 11.3% and 12.88 Coulomb for the current configuration.

The planner will then compose a plan to change the network settings of the current configuration to the settings of the new configuration. Finally, the *Executor* component will use the plan to enact the new network settings to the DeltaIoT system via the management interface. This completes an adaptation cycle.

8.5 Proactive Decision-Making using Probabilistic Model Checking

The two approaches to taming uncertainty presented in the previous sections apply adaptation in response to changes in the short term. This may result in inefficiencies since the system may perform sequences of sub-optimal adaptations. We now look at a proactive approach that makes adaptation decisions with a look-ahead horizon. This approach is based on *proactive latency-aware adaptation*, a well-known adaptation approach that employs formal techniques at runtime to make adaptation decisions. Proactive latency-aware adaptation is inspired by some of the main ideas in model predictive control, which include the use of a receding horizon and the selection of a sequence of control

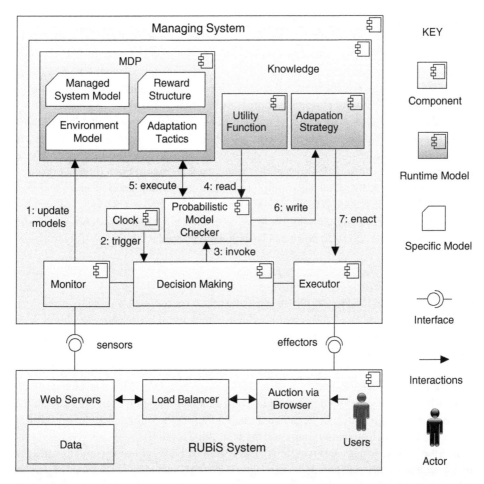

Figure 8.10 Architecture of a feedback loop that uses proactive decision-making applied to RUBiS.

actions from which only the first one is executed, with the sequence being recomputed at the start of the next decision episode.

In short, the central idea of the third approach is to use a formal model of the managed system in which the adaptation decision is left under-specified through nondeterminism. A model checker then resolves the nondeterministic choices so that the accumulated utility over a given horizon is maximized. Figure 8.10 shows the architecture of a self-adaptive system with a feedback loop that applies proactive decision-making using probabilistic model checking.

We study the approach in the context of RUBiS. RUBiS is a Web-based application that consists of a Web tier that receives requests from clients via browsers, a server tier that handles the requests, and a data tier that provides the dynamic content requested by the clients. A load balancer situated between the Web tier and the server tier distributes the requests among the Web servers. The workload on the system depends on the arrival rate of requests, which is the main uncertainty in the system.

The system offers two pairs of tactics to handle changes in the load on the system. The system can add and remove servers. Adding a service takes some time, while removing a service can be done instantaneously. The system also supports a feature that is called brownout. With brownout, responses to requests include mandatory content (data about the item browsed) and possibly optional content (for instance recommendations for users). Processing optional content requires resources. A parameter that is called *dimmer* allows the fraction of responses with optional content to be controlled, hence enabling the system to manage the average number of requests it handles. Here a predefined set of levels are used that discretizes the range of the dimmer parameter. User requests that are efficiently processed, i.e. responded to within a response time requirement, provide revenue for the system owner. Requests with optional content provide additional revenue. The cost to operate the RUBiS system is proportional to the number of servers in use.

The goal of the self-adaptive system is to maximize the difference between revenue and cost. For a duration L, this difference can be expressed using the following utility function

$$U = R_M \cdot x_M + R_O \cdot x_O - C \int_0^L (s(t) - 1)dt \tag{8.6}$$

with U the utility of the system; R_M the revenue of a response with only mandatory content and x_M the number of responses provided within the time window to obtain such revenue; R_O the revenue of a response with optional content and x_O the number of such timely responses; and $s(t)$ the number of services at time t.

The managing system realizes a feedback loop where the analyze and plan functions are integrated into one decision-making component, see Figure 8.10. This component determines whether it is possible to adapt the system to a configuration that will provide the system owner a higher utility (i.e. the analysis function of the self-adaptation work flow), and if so it finds such a configuration (i.e. the planning function).

The *Monitor* component observes the RUBiS system and its environment and updates the corresponding models in the knowledge repository. This includes the number of servers with their status, the current setting of the dimmer, the observed average response time, and the request arrival rate at the load balancer.

The *Decision-Making* component runs periodically, i.e. with interval τ synchronized by an internal clock. This component will invoke the *Probabilistic Model Checker* that will determine whether adaptation is needed and, if so, which set of adaptation tactics offers the highest utility and should be used to adapt the system.

The relevant parts of the self-adaptive system and its environment are represented as a Markov Decision Process runtime model (MDP). The probabilistic model checker uses the MDP to reason about the uncertainty in the environment and find an optimal adaptation strategy based on a reward/cost structure that is determined by the utility function. The MDP specifies how the state of the system can evolve in discrete time steps. In each state s, the system can select from a set of enabled actions $A(s)$; the choice of actions is non-deterministic. When an action is selected, the successor state is chosen based on a probability distribution $\Delta(s, a)$. A strategy resolves the non-deterministic choices of the model by selecting an action in every state based on an objective that can be expressed as a reward-based property. In general, the property $R^r_{max=?}[F^\star \phi]$ enables the maximum accrued reward r along paths that lead to states satisfying the state formula ϕ to be quantified.

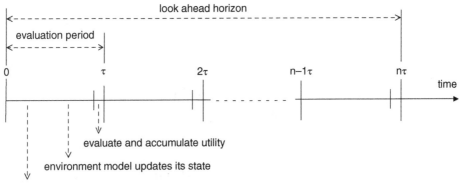

Figure 8.11 Schedule of model execution with probabilistic model checking.

Concretely, the MDP for the RUBiS system that is used by the probabilistic model checker consists of the composition of the *Managed System Model*, the *Environment Model*, the *Reward Structure*, and the *Adaptation Tactics*, see Figure 8.10 and explained in detail below. The schedule of how the composite model is executed is shown in Figure 8.11.

The MDP is evaluated once per period of the wall-clock time τ (corresponding to one invocation of the model checker). The MDP models the evolution of the system and the environment as described in Figure 8.11, where time is advanced over the horizon in the model.

The start of the first evaluation period represents the current time of the system. The evaluation is repeated for n periods that correspond to the look-ahead horizon $n\tau$. In each evaluation period, the managed system has a chance to proactively adapt. Then, the environment updates its state taking a probabilistic transition according to its model. Next, the utility that the system provides for the evaluation period is computed and accumulated. Finally, time is advanced, and the evaluation process is repeated until the end of the look-ahead horizon is reached. The accumulation of the utility will be maximized over the horizon by the model checker using the reward structure.

Making decisions over a look-ahead horizon requires predicting the states of the environment in the near future. To that end, the environment is modeled as a stochastic process that evolves per time step corresponding to the evaluation period τ. The predictions of future states are based on observations of average arrival rates of requests at the load balancer, for which a time series predictor can be used. Figure 8.12 shows a DTMC that models the environment of RUBiS.

The root of the graph S_0 corresponds to the current state of the environment e_0. The other nodes represent estimates of possible future states of the environment conditioned on the parent, with the edges representing the probability that a child state is obtained given that the parent was obtained. As commonly applied in MDP, the model considers three branches with approximations corresponding to the 5th, 50th, and 95th percentiles of the estimation distributions. These estimates are provided by the time series predictor based on the observations of the request arrival rates up until the current state of the environment. The tree is

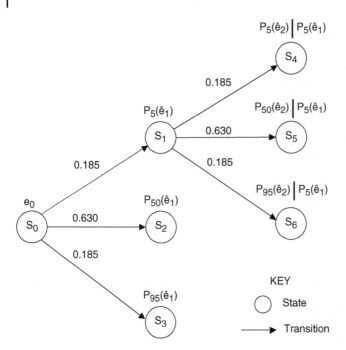

Figure 8.12 MDP that models the environment of RUBiS.

expanded for a number evaluation periods equal to the look-ahead horizon.[2] Given that the state of the environment may change at any time, a new probability tree for the environment is generated at the beginning of the adaptation decision process.

Consider for instance requests made to the website of the 1998 World Cup, a well-known open source benchmark used for the performance evaluation of Web and Cloud based systems. The state of e_0 for this website at some point in time may correspond to a request arrival rate of 10 700 requests per minute. With τ defined as one minute and a look ahead horizon of 3τ, the prediction for $P_5(\hat{e}_1)$ may correspond to 15 800 (burst), for $P_{50}(\hat{e}_1)$ to 10 700 (continuous load), and $P_{95}(\hat{e}_1)$ to 8800 (decreasing load).

In contrast to the environment model that encodes the uncertainty in RUBiS, the models of the managed system and the adaptation tactics do not change at runtime (except for the configuration parameters). The model of the managed system maintains the number of active servers and the current level of the dimmer. This model allows the average response time per evaluation period for a given system configuration and the state of the environment to be computed. When an adaptation tactic is applied, the model will reflect the changes of the state. We illustrate the model of an adaptation tactic for adding a server. This model is structured as follows:

> module *addServer*
>   addServerState: [0...addServer_LATENCY_PERIODS] init *ini_addServerState*;

---

2 In practice, the tree is expanded for a fixed number of levels, and linearization is assumed for the rest of the decision periods to avoid state-space explosion. The effects of this simplification in the results of model checking have shown to be negligible.

addServerGo : bool init *true;*
//tactic applicable start
[addServer_start] addServerGo & addServerState = 0 & addServerApplicable
    → *(addServerState' = 1) & (addServerGo' = false);*
//tactic applicable wait ...
//tactic not applicable ...
//progress tactic ...
//complete tactic ...

The variable *addServerState* keeps track of whether the tactic is executing or not (i.e. whether a server is being added) and if it is executing, how much progress has been made in terms of latency periods. This variable allows decisions to be made consistent with the current state of the managed system and its evolution modeled by the MDP over the look-ahead horizon. At the beginning of each evaluation period, the tactic is evaluated to determine whether it is applicable or not. When the tactic is applicable, different options can apply. Whether the tactic can start depends on three variables: *addServerGo* holds when adding a server is permitted,[3] *addServerState* holds when the tactic is in execution, and *addServerApplicable* holds when the tactic applies in the current situation. When these conditions hold, a server will be added and the condition *addServerGo* set to false. When the tactic is evaluated, it may also have to wait depending on the concrete conditions. When the tactic is in progress, it synchronizes with the model of the managed system (accounting for the latency periods of the executing tactic). The tactic completes when the addition of the server completes. Other tactics (i.e. remove server, increase or reduce dimmer) are defined in a similar way.

Finally, the probabilistic model checker computes the maximum accumulated reward over the look-ahead horizon. This property can be specified in PRISM notation as follows

$$R^{\text{util}}_{\text{max}=?}[F^{\leq C} end] \tag{8.7}$$

The reward structure *util* computes the utility *U* of the system (as defined in the formula given above) at the end of each evaluation period. The predicate *end* indicates the end of the look-ahead horizon. The probabilistic model checker will synthesize a strategy that maximizes the accumulated reward over the horizon. Hence, it will resolve the uncertainties in the composed MDP.

The adaptation tactics selected by the best strategy at the start of the first evaluation period are the tactics that the *Executor* component needs to enact on the managed system. Hence, the decision-making component looks at the optimal strategy over a look-ahead horizon that realizes the highest utility and applies the adaptation tactics at the current time. This process is repeated in every new adaptation cycle. Hence, just the first set of control actions is executed and then the rest of the strategy is discarded. The strategy is fully recomputed at the start of the next decision period, helping the system to cope with potential disturbances that might appear.

As an example, in response to a predicted increase of load, the maximum accumulated reward (utility over the look-ahead horizon) may be achieved by starting by reducing the

---

3 The variable *addServerGo* ensures that the tactic module chooses to start or not only once per modeled period, which is required since each tactic is modeled as a concurrent process.

dimmer value by one level (a fast tactic) combined with adding a new server (a slow tactic). These non-conflicting adaptation tactics can be enacted in parallel right away. Changing the dimmer level enables the system to handle an increase in requests in the short term, until the addition of the new server completes. Depending on the future conditions, the tactic that increases the dimmer level may, for instance, be applied to maximize the accumulated reward.

8.6   A Note on Verification and Validation

It is important to note that the guarantees for the adaptation goals provided by the three representative approaches presented in this chapter hold for the *runtime models of the self-adaptive system* that are used by the feedback loop. The selection of configurations to adapt the managed system relies on the verification of the models of the self-adaptive system during operation. However, since these models are an abstraction of its target that is subject to uncertainty and thus do not necessarily represent the ground truth, the selection of configurations to adapt the managed system is only as good as the runtime models that the feedback loop uses to analyze the adaptation options and make adaptation decisions.

   Verification allows statements to be made about properties of models of the system, which are not necessarily the exact properties of the real system that these models represent. Hence, runtime verification needs to be complemented with validation of *the self-adaptive system itself*. Validation assesses the real system, for instance through testing against the stakeholders' requirements. As such, validation provides evidence for the compliance of the system with its requirements, which is a prerequisite to acceptance of the operational system for its intended use. As such, validation allows the formal guarantees for the predicted compliance of the system with its goals to be approved, which are obtained through verification of the runtime models.

8.7   Integrated Process to Tame Uncertainty

 Uncertainties make it difficult or even impossible to obtain sufficient evidence for requirements compliance of a system before it is deployed. Self-adaptation enhances a system with the ability to collect additional information at runtime to resolve uncertainties and continuously provide new evidence for the the compliance of the system with its requirements. Hence, taming uncertainty in self-adaptive systems calls for an integrated engineering process that combines offline human-centered activities with online machine-centered activities. We focus here on engineering activities related to the managing system only, assuming that the managed system is available and equipped with the necessary probes and effectors to realize self-adaptation.

   Figure 8.13 shows an integrated process for engineering self-adaptive systems to tame uncertainties. The process spans the four main stages of the software life cycle of self-adaptive systems. Stages I and II cover the implementation and enactment of a managing system. Stage III realizes adaptation of the managed system at runtime to achieve the adaptation goals. Stage IV covers the evolution of the managing system to deal with new or changing goals and updates of the managing system.

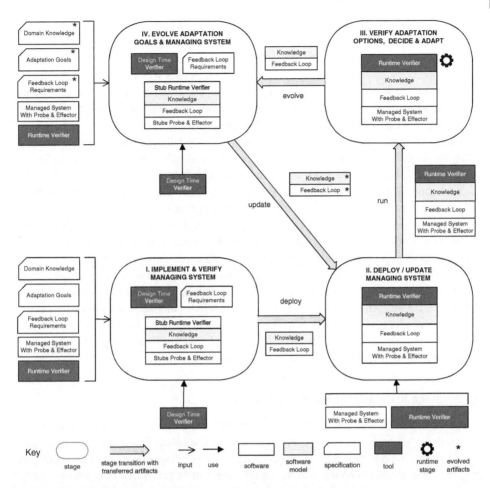

Figure 8.13   The four stages of the integrated process to tame uncertainty.

8.7.1   Stage I: Implement and Verify the Managing System

In the first stage, the feedback loop of the self-adaptive system is developed together with the runtime models of the knowledge. Implementing a feedback loop for a problem at hand requires *domain knowledge*. Domain knowledge refers to domain-specific information provided by stakeholders about the environment and the system itself that is relevant to adaptation. Examples are the expected load of the system, initial values of the uncertainty parameters, and the settings of elements of the system that can be used to reconfigure the system. Furthermore, the developer needs a specification of the *adaptation goals*, which refer to the quality requirements that need to be realized by the feedback loop. The adaptation goals need to be encoded in a machine-readable format that allows the feedback loop model to reason about the goals at runtime.

Additionally, the feedback loop requires models of the managed system and its environment, and parameterized models for relevant qualities. For the managed system and the environment models, the developer uses domain knowledge and the implementation of the *managed system with probes and effectors* to identify the characteristics that are relevant for adaptation. Typically an architectural representation of the managed system and the relevant aspects of the environment in which the system operates are used. The developer uses domain-specific knowledge to specify parameterized quality models for each adaptation goal. On the one hand, the parameters represent uncertainties that need to be monitored and updated at runtime; on the other hand, the parameters represent variable settings of the system that can be used to define the adaptation options.

In order to verify the correctness of the feedback loop behavior, the developer needs a set of *feedback loop requirements* the managing system should comply with. Examples are requirements that ensure the analysis of runtime models is performed correctly or the steps of an adaptation plan are executed correctly. An important requirement is to ensure that a fail-safe configuration is selected to adapt the managed system to under exceptional conditions. Furthermore, verifying the feedback loop requires a set of domain-specific *stubs* that represent the behavior of the external elements the feedback loop interacts with. These stubs are connected to the feedback loop during verification. The stubs can be derived from a specification or from the implementation of the managed system with its probes and effectors and the runtime verifier that will be used to analyze adaptation options at runtime. During the verification of the feedback loop, the knowledge models can either be connected to the feedback loop or these models can be tested separately. The selection of a concrete verification approach depends on the problem and setting at hand, but testing is the dominant approach used in practice, together with model checking.

8.7.2 Stage II: Deploy the Managing System

In the second stage, the verified feedback loop implementation and the knowledge models are deployed and enacted. Deployment includes the instantiation, configuration, and installation of the software of the managing system. Depending on the context at hand, this may require manual intervention or the deployment can be performed fully automatically.

After being deployed, the feedback loop needs to be connected to the managed system, which is realized through probes and effectors. Depending on the concrete realization, the monitor component of the feedback loop can be directly connected to the probes and the executor component with effectors, or the developer may need to provide these links through dedicated code. Next, the *runtime verifier* needs to be deployed and connected to the feedback loop, allowing the analyzer component to estimate the qualities for the adaptation options in order to make decisions about how to adapt the system from its current configuration when needed. This link is typically implemented using dedicated code. Regardless of the type of mechanisms that are used to connect the managing system, it is important that the mechanisms ensure correct communication between the feedback loop and the external elements. When specific classes are developed to realize the connections, such guarantees can be provided through testing.

When the managing system is deployed and the connections are established (with the probes, effectors, and the verifier), the adaptation can be started. Depending on the characteristics of the concrete situation, some configuration may be required before the execution can effectively start. For instance, initial settings of model parameters may need to be loaded and assigned.

8.7.3 Stage III: Verify Adaptation Options, Decide, and Adapt

The third stage is a runtime stage where the executing feedback loop dynamically adapts the managed system. To that end, the feedback loop monitors the managed system and its environment, analyzes the changing conditions, and adapts the managed system to realize the adaptation goals.

The decision-making consists of four basic steps: (i) composing the adaptation options by assigning values to the variables that represent elements of the managed system that can be adapted; (ii) assigning values to the uncertainties as observed by the monitor and stored in the knowledge repository – uncertainties are typically represented as variables in the run-time models; (iii) invoking the runtime verifier to predict the different qualities for each adaptation option; (iv) using the verification results and the adaptation goals to identify the best adaptation option, for which a plan is generated. In the event that the verification process is not able to find a valid adaptation option (step iii), e.g. the current conditions in the environment do not allow such an option, the system should select a fail-safe configuration to adapt the managed system to.

8.7.4 Stage IV: Evolve Adaptation Goals and Managing System

In the fourth stage, the adaptation goals and the managing system evolve based on feedback obtained from the executing system or because the adaptation goals need to be changed. Performing such evolution on-the-fly requires a dedicated infrastructure.

When stakeholders define a new requirement for adaptation, this requirement needs to be translated to an adaptation goal that can be integrated into the decision-making mechanism of the feedback loop. In addition, a new quality model needs to be defined that allows prediction of the quality property that corresponds to each new adaptation goal for the different adaptation options. These domain-specific tasks are similar to the specification of adaptation goals and quality models in stage I. Such evolution typically requires updates of the different components of the feedback loop. The monitor model needs to be extended with support to track data and uncertainties that relate to the new adaptation goal. The analyzer needs to be extended with support to perform analysis of the new quality property. The planner needs to incorporate the new goal into the set of adaptation goals that are used to select the best adaptation option. New types of plan steps may need to be incorporated in the planner, and the executor may need to be extended to deal with new types of adaptation actions. Before updating the running managed system, the evolved managing system needs to be verified. The initial correctness properties can be checked again, possibly complemented by the verification of new domain-specific properties.

When the evolved managing system is verified, it needs to be updated – i.e. the running managing system needs to be replaced with the new implementation. Updating the managing system during operation requires specific infrastructure. This infrastructure should ensure that the managing system is updated while in a safe state to ensure consistency. A typical approach to realize this is to bring the managing system into a *quiescent* state, i.e. a state where no activity is going on in the system. One approach to realize quiescence is by means of using reactive MAPE components, where each component has a dedicated state where it waits to be triggered to start its adaptation function. When all MAPE components are in the waiting state, the feedback loop is in a quiescent state. In addition to support for quiescence, the infrastructure needs to ensure that messages invoked to the feedback loop during an update (e.g. messages with data provided by a probe) are buffered and properly handled after the update. Finally, the infrastructure needs to support the transfer of the state of the managing system after the update and restarting the execution of the feedback loop. Evolving adaptation goals and the managing system typically requires human intervention.

8.8 Summary

Uncertainty in self-adaptive systems is defined as any deviations in deterministic knowledge that may reduce the confidence in adaptation decisions made based on this knowledge.

Uncertainties can be classified into four groups: uncertainties in the system itself (managed and managing), in the goals of a self-adaptive system, in the context of the system, and uncertainty caused by humans involved in the functioning or operation of the self-adaptive system.

Without proper mitigation of uncertainty, the good operation of a software system may be jeopardized. Taming uncertainty refers to providing guarantees for the compliance of a self-adaptive system with its adaptation goals, regardless of the uncertainty it faces.

Uncertainties often make it hard or even impossible to provide these guarantees offline before the system is deployed. Hence, an enduring process is required that provides evidence for the guarantees throughout the lifetime of the system.

A prominent approach to taming uncertainty is the use of formal techniques at runtime. These techniques are used in the decision-making process: on the one hand to analyze the adaptation options and on the other hand to select the best adaptation option.

Runtime quantitative verification is a pioneering technique used to provide guarantees for the adaptation goals of systems with stochastic behavior. The approach usually relies on runtime analysis of Markovian models using properties specified in probabilistic temporal logic. Exhaustive verification suffers from the state explosion problem when the scale of the system increases.

Statistical model checking at runtime is a more efficient technique to provide guarantees for the adaptation goals of a self-adaptive system. This approach allows a tradeoff between the accuracy and confidence of the verification results, and the resources and time required to perform the verification.

Proactive decision-making using probabilistic model checking makes adaptation decisions within a look-ahead horizon. This approach uses a formal model of a self-adaptive system in which the adaptation decisions are left under-specified through non-determinism. A model checker then resolves the non-deterministic choices to find an optimal adaptation decision over a given time horizon.

Guarantees provided by decisions made on runtime models need to be complemented by validation of the real self-adaptive system. Validation allows approval of the formal guarantees for the compliance of the system with its goals, which are obtained through verification of a model of the system.

Taming uncertainty and guaranteeing the compliance of a self-adaptive system requires an integrated process that combines offline human-centered engineering activities with

Table 8.4   Key insights of Wave V: Guarantees under uncertainty

---

- Uncertainty in self-adaptive systems is defined as any deviations in deterministic knowledge that may reduce the confidence in adaptation decisions made based on this knowledge.

- Uncertainty can appear in the system itself, its goals, the context in which it operates, and through humans involved in the system.

- Taming uncertainty refers to providing guarantees for compliance of the system with its adaptation goals, despite the uncertainties the system is exposed to.

- Runtime quantitative verification, statistical model checking, and proactive decision-making are three approaches to taming uncertainty.

- Verification based on models needs to be complemented by validation of the system to provide evidence that the operating system complies with its requirements.

- Taming uncertainty requires an integrated engineering process that spans implementation, enactment, runtime adaptation, and evolution.

---

online machine-centered activities. Such an integrated process spans four stages that cover, respectively, the implementation of a managing system; the enactment of it; its runtime operation where the system realizes adaptation to achieve the adaptation goals; and finally evolution to deal with new and changing goals and updates to the managing system.

Table 8.4 summarizes the key insights of Wave V.

8.9 Exercises

8.1   Sources of uncertainty in self-adaptive mobile applications: level D

Mobile applications are an emerging domain for the application of self-adaptation. A systematic literature review performed in 2019 characterized different modeling dimensions of self-adaptive mobile applications described in 44 primary studies. Collect the papers analyzed in the systematic literature review (possibly extended with new relevant studies), and identify for each of these primary studies the sources of uncertainty considered. Analyze and draw conclusions. For the results of the systematic literature review on self-adaptive mobile applications, see [90]. For background on performing systematic literature reviews, see [118].

8.2   Evaluation of adaptation applied in UNDERSEA: level D

Setting. UNDERSEA is an artifact for engineering a self-adaptive unmanned underwater vehicle (UUV). UNDERSEA comprises a simulated UUV (managed system), and comes with a set of predefined oceanic surveillance missions. In addition, the UNDERSEA distribution offers a predefined adaptation case that performs such missions.

Task. Download and install UNDERSEA. Study the predefined case, which considers adaptation goals for throughput, resource usage, and cost. Pay special attention to the stochastic models of the UUV, which are parameterized with the possible speed and sensor configurations. During the analysis stage of the MAPE work flow, these models are verified using probabilistic model checking at runtime. Also pay attention to the cost function, which expresses the desired trade-off between carrying out the mission with reduced battery usage and completing the mission faster. Run the experiment that comes with the simulator and collect the data. Assess the UUV compliance with the requirements when the system performs runtime adaptations when needed. Now adjust the weights of the cost function and observe the effects. Perform an analysis of the results and draw conclusions.

Additional material. The UNDERSEA artifact can be downloaded via [85]. For more details about the example case, the runtime models, and data collection, see [84]. For additional information about UNDERSEA, see [86].

8.3 Evaluation of adaptation with modified mission in UNDERSEA: level D

Modify the mission in the UNDERSEA simulator by adding/removing sensors and altering their characteristics. Re-execute the mission generation script and observe the effects. Reflect on the results and draw conclusions.

8.4 Proactive decision-making for DingNet: level W

The DingNet artifact supports research on self-adaptation in the domain of IoT. DingNet offers a simulator that maps directly to a physical IoT system that is deployed in the area of Leuven, Belgium. DingNet models a set of geographically distributed gateways that are connected to a user application deployed at a front-end server. The gateways can interact over a LoRaWAN network with local, possibly mobile, motes that can be equipped with sensors and actuators. The DingNet artifact comes with a predefined smart city application where sensors are deployed throughout the city of Leuven that track the quality of the air. The application uses this data to support cyclists in finding paths through the city with good air quality. Download the DingNet artifact and experiment with it. Take a predefined configuration and test it using the standard solution that determines paths using A*. Now design a solution that applies proactive decision-making to determine paths for cyclists. Implement your solution and compare it with A*. Make a tradeoff analysis. The DingNet artifact can be downloaded via [203]. For additional information about DingNet, see [158].

8.5 Feedback loop model for simple IoT network: level D

Setting. Consider the simple IoT network that we used in Exercise 6.5. Copy the Uppaal model of the simple IoT network from the DeltaIoT website [213] and save it as an XML file. Open the model with Uppaal. This model specifies the state and the behavior of the simple IoT network over 10 cycles. The IoT system is equipped with a probe and an effector. The probe enables monitoring the current state of the IoT system, while the effector allows the network settings to be adapted. The patterns for the messages generated by Mote [2] and the interference in the environment are defined as two arrays with arbitrary selected phases (see Exercise 6.5). Furthermore, the model of the simple IoT network provides an abstract definition of the managing system comprising abstract templates for the MAPE functions and a specification of a skeleton model of the knowledge as shown in Figure 8.14.

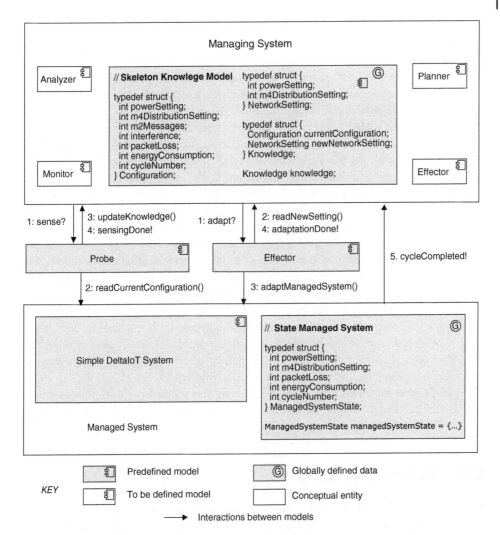

Figure 8.14 Model of the simple IoT network.

The probe can be triggered via the *sense* channel. The probe will then read the following data of the IoT system: the current power setting of the motes, the current setting of the distribution of messages used by Mote [4], the number of messages generated by Mote [2] in the last cycle, the current level of interference, the packet loss of the network in the last cycle, the energy consumed by the network in the last cycle, and the current cycle number (starting from 1). The probe writes this data to the knowledge model and signals the end of the sense action via the *sensingDone* channel. The updated data of the knowledge model can then be read by the MAPE elements to make adaptation decisions. To adapt the network settings, the MAPE elements need to provide new values of the power setting and the distribution of the messages applied by Mote [4]. When new settings are written to *newNetworkSettings* of the knowledge model, the effector can be triggered via the *adapt* channel. The effector will then read the new network settings from the knowledge model, adapt the IoT system accordingly, and signal the end of the adaptation action via the *adapationDone* channel. When the managed system ends its current cycle, it will signal the managing system via the *cycleCompleted* channel. The Uppaal model provides a simple test template that illustrates the interactions with the probe and effector for one adaptation cycle. You can simulate this test model with the Uppaal simulator.

Task. Design a MAPE-K feedback loop that keeps the average packet loss over 10 cycles below 10%. To that end, instantiate the abstractly defined MAPE templates and extend the skeleton knowledge model as needed. For the analysis of the different adaptation options, apply simulation. To that end, you can reuse the quality model for packet loss designed in Exercise 6.5. Select a particular system configuration and apply one adaptation cycle using the Uppaal simulator. Select a few alternative configurations and repeat. Observe the effects on the packet loss goal. Then evaluate your solution for 10 cycles and assess your solution. Now extend your MAPE-K feedback loop so that it also minimizes the energy consumption over 10 cycles. For the analysis of energy consumption of different adaptation options, you can reuse the corresponding quality model designed in Exercise 6.5. Evaluate your solution first for one adaptation cycle using the Uppaal simulator. Then apply 10 cycles. Critically assess your solution with respect to the following aspects: How optimal is your design? How would your design be affected if another third adaptation goal were added? How reusable is your solution if the uncertainty patterns changed? What are possible options to transfer your design into a practical implementation?

Additional material. The Uppaal tool can be downloaded via [189]. Reusable templates that may be helpful for the design of the different MAPE models can be found in [104] and [202].

8.10 Bibliographic Notes

D. Perez-Palacin and R. Mirandola classified uncertainties in self-adaptive systems in three dimensions: location, level, and nature [155]. Location refers to the place where the uncertainty manifests itself, level refers to a value on the scale from deterministic knowledge to total ignorance, and nature refers to whether the uncertainty results from imperfection in the acquired knowledge or from inherent variability in the phenomena under consideration. S. Mahdavi-Hezavehi et al. conducted a systematic literature review on uncertainty in architecture-based self-adaptive systems [137]. From the data extracted from 51 primary studies, the authors propose a classification framework for uncertainty and its sources. Esfahani et al. emphasized uncertainty in the environment (or domain) in which the software is deployed as a prevalent aspect of self-adaptive systems [69].

Early research on the use of formal verification techniques in adaptive systems concentrated on ensuring correctness by construction of an adaptive system. A classic example is described in [219] where an adaptive system is modeled as a Petri net, which is used to verify a set of invariants and integrity constraints. For an earlier survey on the use of formal verification techniques in self-adaptive systems, see [205]. R. Calinescu et al. argued for the use of quantitative verification at runtime to support the decision-making in self-adaptive systems [40]. The concept of perpetual assurances for self-adaptive systems that was coined in [210] refers to an enduring assurance process in which humans and the system jointly derive and integrate new evidence and arguments, spanning the whole lifetime of the system.

R. Calinescu et al. introduced the QoSMOS approach, where a service-based system is modeled as a Markov model, and adaptation goals are expressed as probabilistic temporal logic formulae. The properties are automatically analyzed at runtime to identify and enforce optimal system configurations [39]. PRISM is described in [123], and the tool can be downloaded via [156]. C. Baier and J. Katoen explain the principles of model checking in [10].

U. Iftikhar et al. introduced ActivFORMS, an approach that applies statistical model checking to automatically analyze the adaptation options of a self-adaptive system at runtime [101, 202]. R. Alur and D. Dill described the basic theory of timed automata [3]. A. David et al. presented a tutorial on Uppaal that includes support for analyzing networks with stochastic semantics [57]. The Uppaal tool can be downloaded via [189].

G. Morena et al. introduced proactive latency-aware adaptation [144] that uses a formal model of the adaptive system in which the adaptation decision is left underspecified through nondeterminism, and has the model checker resolve the nondeterministic choices so that the accumulated utility over the horizon is maximized. The RUBiS artifact can be downloaded via [168], where plenty of documentation is available about how to use the artifact.

J. Andersson et al. emphasized the need for systematic engineering processes for self-adaptive systems [6]. The integrated process to tame uncertainty described in this chapter is based on ENTRUST [42], an end-to-end methodology for the engineering of trustworthy self-adaptive software, extended with support for online evolution based on ActivFORMS [101, 202].

9

Wave VI: Control-based Software Adaptation

The sixth wave is centered on the application of control theory as a basis to realize and analyze self-adaptive software systems that operate under uncertainty. The second wave, Architecture-based Adaptation, laid a basis for the systematic engineering of self-adaptive systems, but raised the need for a theoretical foundation for self-adaptation. The fifth wave, Guarantees Under Uncertainty, tackled the problem of taming uncertainty, but the solutions developed to provide guarantees for the compliance of a system with its adaptation goals were shown to be complex. The insights obtained from waves two and five were the drivers for the sixth wave, which focuses on exploiting the mathematical foundation of control theory for realizing self-adaptive systems and analyzing these systems to guarantee their key properties.

Control theory is a mathematically-founded discipline that dates back a century; it provides techniques and tools to design and formally analyze feedback loop systems. The basic idea of control theory is relatively simple: an external controller is added to a system to form a feedback loop. The controller monitors a variable of the system (which is subject to disturbances), compares its value with a reference value (the system goal), and based on the observed difference generates a control action that is enacted on the system to bring the variable to the reference value. A major benefit of control theory is that it allows various properties of feedback loops to be formally analyzed, for instance their stability and accuracy to achieve the objectives within bounded settling times. Starting from this basic idea, a broad range of control strategies have been developed, each with specific characteristics and properties. The application of feedback control has been particularly successful in a wide range of domains ranging from automotive and aircraft to manufacturing and industrial processes.

Over a decade ago, researchers were already arguing that control theory could play a central role in managing computing systems. Since then, different control solutions have been developed and successfully applied to various problems of computing systems, including network protocols, data centers, and more recently the Cloud. However, a number of studies have shown that most of the proposed approaches tend to solve domain-specific problems. In addition, the focus of these efforts has mainly been on controlling lower-level elements and resources of computing systems, such as CPU, storage, and bandwidth. Applying control theory to adapt software at higher levels of the technology stack is a more complex

An Introduction to Self-Adaptive Systems: A Contemporary Software Engineering Perspective,
First Edition. Danny Weyns.
© 2021 John Wiley & Sons Ltd. Published 2021 by John Wiley & Sons Ltd.

problem. One of the main reasons that has been put forward for this complexity is the difficulty of accurately modeling software, which often has non-linear behavior. For instance, the utilization of a pool of services is typically not proportional to the actual load on the system. Other reasons include the difficulty of instrumenting software to obtain measurements from sensors and enact control to the system through actuators, the diversity of quality requirements of software, and their complex interplay. Last but not least, a factor that hampers control-based adaptation of software is the complexity of the mathematics needed for software engineers to apply control techniques.

In response to these problems, researchers aimed at developing control-based solutions that *automatically* construct a dynamic model of a software system with a suitable controller for managing the system's quality requirements, while preserving the power of the guarantees provided by control theory. The basic control strategy used in automatically constructed control-based solutions for software adaptation is *feedback control*. Feedback control measures the system output and adjusts the system to meet a desired output response in spite of the disturbances the system is exposed to. In contrast, *feed forward control* only takes into account the goal and the external disturbances, and produces a control signal that compensates for the disturbances.[1]

We start this chapter with a brief introduction of the basics of control theory. Then we explain three characteristic approaches that apply control-based software adaptation. The first approach, which laid the foundations of automating the construction of control systems, relies on Proportional-Integral control to handle a single setpoint goal. The second approach extends the first approach and handles multiple goals, including an optimization goal. The third approach applies Model Predictive Control to make optimal control decisions over a look-ahead horizon for multiple goals.

LEARNING OUTCOMES

- To motivate the application of control theory to realize self-adaptive software systems.
- To explain the block diagram of a feedback control loop and discuss its main properties.
- To explain the main control properties that apply for self-adaptive software.
- To explain the phases of automatic controller construction and operation.
- To motivate and explain online updates of the system model.
- To explain and illustrate the principles of an automatically constructed Single-Input Single-Output (SISO) control system.
- To explain and illustrate the principles of an automatically constructed Multiple-Input Multiple-Output (MIMO) control system.
- To motivate the need for Model Predictive Control in self-adaptive software systems.
- To explain and illustrate the principles of an automatically constructed Model Predictive Control (MPC) system.

---

1 In principle, feed-forward control can also just use a setpoint value, but this is rarely used in practice.

9.1   A Brief Introduction to Control Theory

Figure 9.1 shows a block diagram of a basic feedback control loop.

In essence, a control-based computing system consists of a target system that is subject to adaptation and a controller that implements a particular control strategy to adapt the target system. The task of the controller is to ensure that the output of the system is as close as possible to the reference input, while reducing the effects of uncertainty that appear as disturbances, noise in variables, or imperfections in the models of the system or environment used to design the controller.

Concretely, the basic elements of a feedback control loop are:

- Target system (also called Plant): the system that is manipulated by a controller to achieve the desired output in the presence of disturbances.
- Controller: the external element that is added to the target system to dynamically adjust its behavior based on the difference between the measured system output and the reference input.
- Transducer: the element that transforms the measured output so that it can be compared with the reference input (e.g. smoothing noise of the measured output).

The elements use the following signals, which are functions of time (represented by k):

- The reference input $r(k)$: the desired value of the measured output at time k.
- The control error $e(k)$: the difference between the reference input and the measured output.
- The control signal $u(k)$: the parameter setting that allows the behavior of the target system to be dynamically adjusted.
- The disturbance input $d(k)$: any exogenous phenomena that interfere with the effects of the control input on the measured output.
- The measured output $y(k)$: the measurable parameter of the target system.

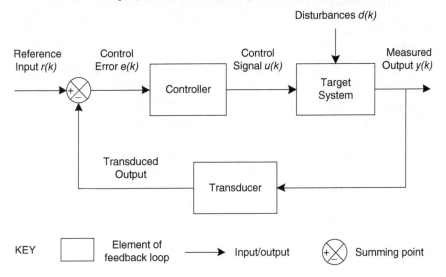

Figure 9.1   Block diagram of a basic feedback control loop.

The target system corresponds to the managed system in the reference model of a self-adaptive system. The reference input corresponds to the adaptation goal, and the controller realizes a feedback loop.

There are three main purposes for control. *Regulatory control* aims to keep the measured output of the target system equal to (or near) the reference input. *Disturbance rejection* aims to suppress the effects of disturbances that act on the system on the measured output. Finally *Optimization* aims to find the best possible value of the measured output, for instance minimizing or maximizing a quality property of the system.

9.1.1 Controller Design

Using control theory to build self-adaptive software requires a model of the target system in the form of a dynamic system. In general, this model defines the relationship between effector settings that can be used to control the system (input variables), state variables and internal dynamics, and the control goals (output variables). In control theory, the model of the target is system is analytic and described mathematically.[2] A system model can be specified manually based on knowledge of the target system, or the model can be identified based on experiments (or a combination of the two). The dynamics of the system can be specified in continuous or discrete time; the latter is the one mostly used for controlling software systems. In general, in discrete time there is a distinction between equidistant sampling, where time instants k are equally spaced with a certain period, and non-uniform sampling, where samples are not taken equally spaced in time, but depend for instance on a function that represents the activation of a software element. A variety of model types can be used to specify the target system, ranging from Markovian models and queuing models to a set of difference equations. For instance, the input variable of a system with a queue may be the number of incoming requests, the output variable may be the average service time, and the state variable may be the number of requests in the queue. If difference equations are used, the evolution of a discrete time system can in general be written as follows

$$x_{k+1} = f(x_k, u_k) \tag{9.1}$$
$$y_k = h(x_k, u_k) \tag{9.2}$$

where x_k is the state of the system at time instance k ($k = 0, 1, 2, \ldots$), u_k is the input, and y_k is the output. These difference equations tell us how the state x_{k+1} differs from the state x_k and how the output y_k relates to state x_k and the input u_k. Often, for software systems, the system behavior is approximated, and equations can be used that are linear in the state and input.

To design the controller, a variety of techniques can be used that differ in the information that is required to develop a control strategy and the guarantees that they can offer. We briefly illustrate the basics of Proportional-Integral (PI) control, a popular controller in general and also commonly used to control software systems. The value of the control signal generated by a PI controller that is used to manipulate the target system is defined as follows

$$u(k) = K_P.e(k) + K_I \cdot \sum_{i=1}^{n_k} e_i(k)\Delta t \tag{9.3}$$

with $u(k)$ the value of control signal at time k; K_P the proportional gain; $e(k)$ the time varying value of the control error equal to PV-SP, with PV the measured process variable

---

2 Besides analytic models, controllers can also work with grey box and black box models.

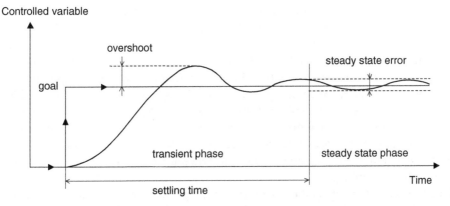

Figure 9.2 Overview of control properties for a response to a step change in the goal.

(measured output) and SP the reference input (set point); K_I the integral gain; Δt the time between sampling instances; and n_k the number of sampling instances.

Whereas the proportional term considers only the current error at the time of the controller calculation, the integral term continually sums up error: it considers the history of the error $e_i(k)$ with $i = 1 \dots n_k$, i.e. how long and how far the measured variable has been from the goal over time. Hence, the contribution from the integral term of the controller is proportional to both the magnitude of the error and the duration of the error.

The values of K_P and K_I are two tuning parameters that determine how the controller brings the control variable to the reference input in response to external changes, including disturbances and goal changes. The proportional term K_P is used to increase the speed of response, while the integral term K_I is used to eliminate any steady-state error when the control variable converges to the reference input.

9.1.2 Control Properties

The main properties that can be analyzed in control theory are stability, accuracy, settling time, and overshoot – sometimes referred to as the SASO properties. Figure 9.2 shows a graphical overview of the main control properties for a response to a step change in the goal.

A system is stable if for any bounded input, the output is also bounded. Stability is a basic property of a control feedback loop system. Applied to a software system, stability implies that the output of feedback control (controlled variable) converges to an equilibrium value after a change, although this may not be a constant value due to the stochastic nature of the system. In particular, a software system is stable if it operates in a region where it performs as required (for instance, achieves a required throughput determined by workloads and settings of the system). The utility of an unstable system is very limited; for instance, response times may be high and variable. Hence, stability is a prerequisite for considering the remaining SASO properties.

Accuracy refers to the convergence of the measured output to the goal (or the optimal value for an optimization objective). Accuracy ensures that the control goals are met (for instance ensuring that the required throughput is achieved without exceeding response time constraints). Accuracy is usually not directly quantified, but measured. For a system in steady state, its inaccuracy, or steady-state error, is the steady-state value of the control error $e(k)$. The steady-state error refers to the difference between the steady state value the system converges to and the desired goal.

The settling time of a control system expresses the time it takes for the controller to converge to the steady-state value. Settling time expresses how quickly the system reaches its goals, so it accounts for the duration of the transient phase, i.e. the time to converge within an acceptable distance of the steady-state value. Convergence in many systems may be asymptotic (i.e. in the limit) or only within predefined bounds (bounded stability). Settling time is particularly relevant when the controlled system has to react to a sudden change in environmental conditions (disturbances) or changes in the value of the goal. However, while a fast reaction to changes is a desirable property, it may result in the controlled system following noise or transitory variations in the disturbances. Most controllers require the user to decide on a suitable trade-off between settling time and noise rejection, and tune them accordingly.

Finally, overshoot refers to the maximum difference between the measured value and the goal during the transient phase. Usually, the system should achieve its goal with limited overshoot as overshoot increases variability in the measured output, which may cause temporary violations of systems requirements.

An additional inherent control property is robustness, which refers to the ability of a control system to perform correctly despite imprecision and partial information in the model describing the system and the assumptions about the intensity and dynamics of its disturbances. Phrasing it differently, robustness is the ability of a closed loop system to be insensitive to variations, such as imprecisions or variations in the model used by the controller, or inaccurate or delayed measurements. Robustness makes it possible to design a feedback system based on (strongly) simplified models.

Control properties of feedback loops are used in two ways. On the one hand, the properties are used to assess the system, i.e. the properties can be guaranteed analytically given the mathematical models of the system and the controller connected in the feedback loop. Thus, a self-adaptive software system that relies on a feedback control loop can give stakeholders guarantees on its convergence to the goal, the time it takes to converge, and how robust it is to disturbances and inaccuracies. On the other hand, the properties can serve as design objectives. In this case, a feedback loop system is designed such that it is stable and has acceptable values of steady-state error, settling time, and maximum overshoot.

9.1.3 SISO and MIMO Control Systems

Figure 9.1 shows a basic structure of a control system with single instances of each of the constituent elements of the architecture. However, in practice, multiple instances may be used. Consider for instance a Web service system (target system) for which we want to keep the number of failed service invocations below a required threshold, for instance 0.5% (reference input). To that end, a controller is added to each server that tracks the failure rates of service invocations (measured output) and controls the selection of service instances based on their actual failure rates (control signal) to control the number of service failures. Disturbances include the arrival rates of service requests and the types of requests, which can change dynamically. Control theory allows the distribution of service instances to be determined for the different types of services so that the resulting system is stable and adapts in a timely and smooth manner in response to disturbances.

Furthermore, the basic control schema in Figure 9.1 corresponds to a single-input single-output (SISO) control system, where one output is controlled by one control signal. In practice, multiple-input multiple-output (MIMO) control systems may be used, where a number of output variables are controlled by a number of control signals.[3] The presence of multiple inputs and multiple outputs requires the controller to handle multiple, possibly conflicting, goals and selecting the most appropriate control signal vector when multiple such vectors can achieve the target goals. Typically, the cardinality of the control signal vector is expected to be greater than or equal to the cardinality of the measured output vector, with at least one control signal affecting each of the output signals. In the example with a Web service system, an additional goal that is usually required from the system is to ensure that the cost of service invocations remains under a certain threshold (reference input). To that end, the controller tracks the cost of service invocations (measured output) and controls the selection of service instances based on their actual cost (control signal) to adapt the cost for using the service. Note that the controller now needs to select service instances such that both low failure rate and low cost are ensured, which may, in general, be conflicting goals.

9.1.4 Adaptive Control

Particularly interesting for controlling computing systems is *adaptive control,* which adds an additional control loop for adjusting the controller itself, typically to cope with slowly occurring changes of the controlled system, or to compensate for possible inaccuracies in the initial system model by adjusting the controller's action based on measurements of the system collected at runtime. For instance, in the example of the Web service system, the basic controller reacts to bursts of service requests and sudden changes in the types of service requests to control the failure rate and cost of service invocations. A second controller may use recursive linear regression to update some of the parameters of the system model used by the first controller during execution in order to provide a more precise representation of the changing context. This will allow the basic controller to use up-to-date information when making its control decisions.

9.2 Automatic Construction of SISO Controllers

The need for reusable control-theoretical solutions for self-adaptive software on the one hand and the lack of deep knowledge of control theory in software engineers on the other hand have led to the development of approaches that allow controllers to be constructed *automatically.* These approaches rely on the following assumptions:

1. The requirements that need to be satisfied by the control system are known; these requirements can be translated into quantifiable goals.
2. The target software system (i.e. the managing system) is available and can be used to run experiments.
3. The target system provides a set of sensors that can be used to measure whether the goals are satisfied.
4. The target system provides a set of actuators that can be used to modify the system in order to realize the goals.

Under these assumptions an approximate model of the software system can be built automatically. This model can then be used to construct a controller that controls the target system in order to achieve its goals.

---

3 For MIMO control systems, each of the signals in Figure 9.1 represents a higher-dimensional vector.

In this section, we present a basic approach to automatically constructing a control system with a single goal and a single actuator. This approach is based on the so called Push-Button Methodology, a pioneering approach to automating the construction of controllers for self-adaptive software. Despite its generality, this approach provides formal guarantees for the dynamic behavior of the system.

9.2.1 Phases of Controller Construction and Operation

Figure 9.3 shows a schematic diagram of a SISO control system with the phases of its automatic construction and operation.

The *model building* phase (switch S in position ①) automatically constructs a linear model of the target system. The model is identified by running on-the-fly experiments on the software of the target system. In particular, the system is tested using a set of systematically sampled values of the control signal, and the effects on the output are measured. Figure 9.5 illustrates the model building phase.

The vector U of control signals used for identification, which needs to be provided by the designer, is

$$U = [min, min + \delta, min + 2\delta, min + 3\delta, ..., max] \tag{9.4}$$

where *min* and *max* are the minimal and maximum achievable values for the goal. δ is the sampling rate, which is a tuneable parameter that is usually set to $\delta = (max - min) \times 0.05$. A higher sampling rate provides a more accurate model but increases the identification time; the choice of δ is domain-specific.[4] By applying regression on the results, a system model is

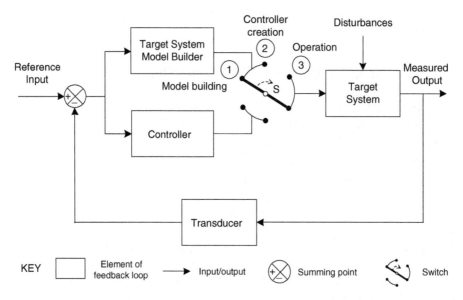

Figure 9.3 SISO control system with its construction and operation phases.

---

4 Here we apply simple equidistant sampling. A more advanced sample distribution concentrating more samples around the regions of the input domain where the system may exhibit faster dynamics may be

Figure 9.4 Illustration of model building.

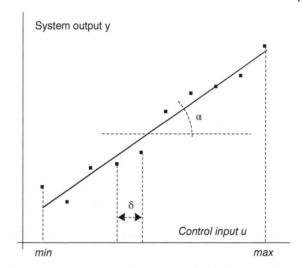

generated in the form

$$y(k) = \alpha.u(k-1) \tag{9.5}$$

where $y(k)$ is the measured output at time instance k, $u(k-1)$ the control signal at the previous time instance, and α the slope of the linear function that gives the rate of change of the output.

The results of the first phase are then used in the *controller creation* phase (switch S in position ②) to automatically create a controller. The goal of this controller is to keep the output close to the reference input while rejecting disturbances. The controller has the form

$$u(k) = u(k-1) - \frac{1-p}{\alpha}.e(k) \tag{9.6}$$

The controller selects a control signal $u(k)$ based on the previous value $u(k-1)$ and the error between the desired reference value and the measured value $e(k) = r(k) - y(k)$. α is directly derived from the system model, and p represents the pole of the transfer function of the control system, which determines the properties of the controller; we explain this below.

Finally, in the *operation* phase (switch S in position ③), the controller exercises its control on the target system, keeping the output as close as possible to the reference input.

9.2.2 Model Updates

The linear model that is constructed during the model building phase is an approximated representation of the target system, which does not capture variations of the system that may occur in practice. To anticipate potential errors in the model, different techniques can be applied. We look here at two standard techniques used in the application of control theory to adapting software systems during operation: incremental model updating and model rebuilding, illustrated in Figure 9.5.

---

useful. However, this would require additional knowledge from the designer, which equidistant sampling does not require.

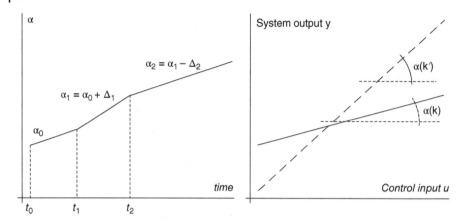

Figure 9.5 Illustration of incremental model updating (left) and model rebuilding (right).

Incremental model updating aims at dealing with relatively small variations of the target system that occur slowly; an example is a gradual change between work modes that are characterized by different loads on the target system. Incremental model updating is real-ized using a mechanism that tracks the response of the system and automatically adjusts the value of the slope of the model (α) to deal with changes in the system's dynamics. This is illustrated in the left diagram of Figure 9.5. At t_0 a model of the target system is used with a slope α_0. At t_1 a variation of the target system is observed and the slope of the model is slightly increased with a value Δ_1. At t_1 the model is updated again; this time the slope is slightly decreased with a value Δ_2. A common mechanism that is used to estimate changes of α is a Kalman filter. Kalman filtering is based on an algorithm that takes a series of mea-surements over time, in our case measurements of the output of the target system, and produces a likely more precise estimate of a variable, in our case the slope of the model of the target system to be used for computing the next control signal. Kalman filters prove to be robust to noisy measurements, achieving more accurate estimations efficiently. The estimated value of $\alpha(k)$ at time k is then substituted in the controller equation so that the controller is working with an up-to-date model. Incremental model updating allows some nonlinear dynamics to be dealt with by updating the slope parameter to obtain a local lin-earization of the system model around the current operation point, as well as capturing small variations in the system's first-order dynamics (i.e. captured by the slope parameter, which estimates the first derivative of the measure with respect to the control input). Note that adding an incremental model updating mechanism on top of a controller results in an adaptive controller that modifies the controller's parameter depending on the variations of the target system.

Model rebuilding aims at dealing with abrupt variations in the conditions of the target system, which are expected to happen rarely. An example is a failure of some core elements of the system. Model rebuilding is realized using a change point detection mechanism that identifies when abrupt change has occurred. This will then trigger a restart of model build-ing followed by controller creation. This is illustrated in the right diagram of Figure 9.5. At time instance k a model of the system is identified that has a slope $\alpha(k)$. At time instance

k' an abrupt change has occurred, and a new system model is identified with a slope $\alpha(k')$. A simple realization of a change point detection mechanism works over a time window of n control actions and computes the average error e_1 for the first $n/2$ samples and the the average error e_2 for the remaining samples. If $| e_1 - e_2 |$ surpasses a given *threshold*, model rebuilding is triggered. As a result, a new model of the target system will be built from scratch, and the resulting value for α will be used to create a new controller that is adapted to deal with the changed operating conditions. The values of n and *threshold* determine when model building will be triggered. It is important to note that model rebuilding is an invasive procedure and should be used only under exceptional circumstances.

9.2.3 Formal Guarantees

Formal assessment allows the stability, steady-state error, settling time, and overshoot of a control system to be analyzed. Such analysis is commonly performed in the Z-domain. A Z-transform converts a representation in discrete-time, where time is represented by a sequence of k discrete instants, into a frequency-domain representation, where z represents the frequency and $1/z$ a one-step temporal delay between actuation and its effect.

For the closed feedback loop system as shown in Figure 9.6 (where the output is directly used to compute the error, compare with Figure 9.1), we can define the following relations in the Z-domain

$$Y(z) = P(z).U(z); \ \ U(z) = C(z).E(z); \ \ and \ E(z) = R(z) - Y(z) \tag{9.7}$$

with $P(z)$ representing the target system (i.e. its Z-transform), $C(z)$ the controller, $Y(z)$ the output of the target system, $U(z)$ the control signal, $E(z)$ the error, and $R(z)$ the reference input.

Solving $Y(z)$ in terms of $R(z)$ gives

$$Y(z) = \frac{P(z).C(z)}{1 + P(z).C(z)}.R(z) \tag{9.8}$$

The transfer function of the closed feedback loop system is then defined as follows

$$G(z) = \frac{Y(z)}{R(z)} = \frac{P(z).C(z)}{1 + P(z).C(z)} \tag{9.9}$$

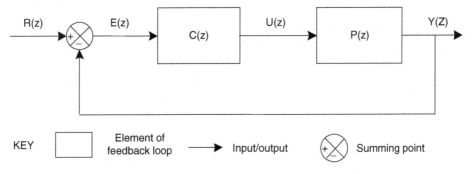

Figure 9.6 Closed feedback loop system.

where $G(z)$ represents the relationship between the control output $Y(z)$ and the reference input $R(z)$ in general.

For the automatically constructed SISO control system, we consider a closed loop transfer function $G(z)$ of the form

$$G(z) = \frac{1-p}{z-p} \qquad (9.10)$$

where p is the pole of the transfer function, which determines the dynamic behavior of the control system. The static gain of the transfer function is 1, meaning that the reference input is directly transferred to the output, which we obtain for $z = 1$.

Substituting the closed loop transfer function $G(z)$ in Equation 9.9, solving this for $C(z)$, and transforming the result to the time domain using the Z inverse transform, we can obtain the controller function

$$y(k) = r(k).(1 - p^k) \qquad (9.11)$$

This function provides an analytical representation of the recursive relationship between measured output and the reference input (setpoint) of the feedback loop in Figure 9.6. The function says that with increasing time k, the system output $y(k)$ converges to the reference input $r(k)$ at an exponential rate, given that $0 \leq p < 1$.

According to control theory, for stability, we require $0 \leq p < 1.^5$ The pole p is a parameter of the controller that can be set by the designer (see the controller function of Equation 9.6).

We define the settling time as the time it takes for the system output to reach $(100 - \epsilon)\%$ of the reference value $r(k)$. We can then rewrite the output as $y(k) = (1 - 0.01\epsilon) . r(k)$, meaning that the system output is within ϵ of the goal. The transfer function in the steady state phase is then

$$(1 - 0.01\epsilon) . r(k) = r(k) . (1 - p^k) \qquad (9.12)$$

The time instance k_ϵ when the output enters the steady state phase is

$$k_\epsilon = \frac{\log\ 0.01\epsilon}{\log |p|} \qquad (9.13)$$

with $\epsilon = 5\%$, the value of k_ϵ is

$$k_\epsilon = \frac{\log\ 0.05}{\log |p|} \qquad (9.14)$$

This value only depends on the value of the pole p. Hence, the pole not only determines whether the control system is stable, but also how fast it reaches the steady state phase. Note that there is a tradeoff between settling time and overshoot. From Equation 9.11 it can be observed that smaller values of p lead to a faster convergence ($1 - p^k$ decreases more rapidly with k), while values of p close to 1 lead to a slower convergence. On the other hand, smaller values of p may lead to more overshoot, while higher values of p result in better disturbance rejection.

---

5 More precisely, p is a complex number that should be inside the unit circle. However, both for complex numbers and for real numbers that are between -1 and 0, the resulting closed loop behavior would oscillate before converging to the one single value that it tends to. Hence, we restrict p to be real valued and between 0 and 1.

9.2.4 Example: Geo-Localization Service

We illustrate the approach with an online geo-localization service. The service provider wants to provide the service with a reliability $r(k)$, i.e. the probability that a service invocation fails should not exceed $1 - r(k)$. The service can be provided on two types of map services, $S1$ and $S2$, which provide different levels of reliability that may change at runtime due to changing workloads and network timeouts.

We enhance the system with an automatically constructed SISO controller that needs to decide for each incoming request which alternative map service to select to ensure the required reliability of the geo-localization service.

The service-based software system offers a monitor that estimates at each point in time the actual reliability $y(k)$ of the geo-localization service. This enables the control system to determine the error $e(k) = r(k) - y(k)$. Furthermore, the software system offers an actuator that allows the probability $u(k)$ that $S1$ is selected to be set (hence the probability that $S2$ is selected is $1 - u(k)$).

Figure 9.7 shows experimental results of the SISO controller realization.[6]

In this scenario, the load on the system may fluctuate and network timeouts may happen, but we do not consider catastrophic events. The system starts with a reliability of 0.5 for service $S1$ and 1.0 for $S2$. The model building phase is marked with gray shading (k from 0 to 400). During model building the probability u is systematically changed from 1 to 0. When the model building phase completes a controller is created that achieves the reliability goal of 0.75. To that end, service $S1$ is selected with a probability around 0.6 (and service $S2$ with a probability around 0.4). At time step 1000, the reliability goal is changed to 0.65. The graphs show the step response of the controller to this change. The results show that the controller is stable and adjusts the probability settings without overshoot. Between steps 1400 and 1700, the reliability of service $S1$ smoothly increases from 0.50 to 0.6. During this period, the Kalman filter identifies the changes and starts to dynamically adjust the slope α of the model of the geo-localization software. As a result, the system converges to a probability setting of $u = 0.5$.

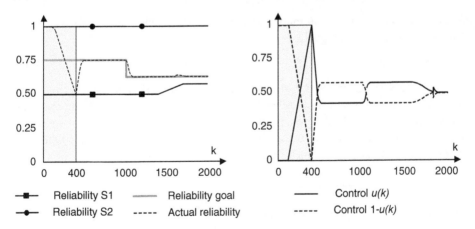

Figure 9.7 Experimental results of the SISO controller realization.

---

6 We define reliability here as 1 minus the failure probability.

9.3 Automatic Construction of MIMO Controllers

The basic approach to automatic construction of a controller for self-adaptive software presented in the previous section is limited to a single adaptation goal and a single control signal. This is often too restrictive for real-world applications. For instance, users of the geo-location service not only want the service to be reliable, but may also want fast response times, while the cost for using the service should not exceed an agreed fee. Such requirements require an approach for the automatic construction of a controller that satisfies multiple goals, often including an optimization goal. This section presents such an approach, which realizes a control schema with multiple inputs that control multiple outputs (MIMO) that can handle multiple goals. The approach builds upon the principles of the automated approach for constructing SISO controllers and hence relies on the same assumptions (listed in Section 9.2). The approach is based on SIMCA, one of the pioneering approaches to automatically generating controllers with multiple goals.

9.3.1 Phases of Controller Construction and Operation

The three phases of the basic approach to automatically constructing a controller for self-adaptive software apply here as well, but they need to be refined for different goals and extended to deal with an optimization goal.

During the model building phase, a set of n linear models of the controlled system is automatically built, where n is the number of goals excluding the optimization goal. Each model M_i is responsible for one goal r_i.

Similar to the basic approach, identification starts by systematically feeding sampled values of goal r_i in the form of a control signal u_i to the target system and measuring the effect on the system output y_i. The vector U_i of control signals used for the identification of the model for goal r_i is

$$U_i = [min_i, min_i + \delta, min_i + 2\delta, min_i + 3\delta, ..., max] \qquad (9.15)$$

where min_i and max_i are the minimum and maximum achievable values for goal r_i.

The resulting model M_i for goal r_i is

$$y_i(k) = \alpha_i.u_i(k-1) \qquad (9.16)$$

The model describes the relationship between the output $y_i(k)$ and the control signal $u_i(k-1)$ for goal r_i, with α_i the slope of the linear function of the system model for that goal.

The system models for the different goals are then used in the *controller creation* phase to automatically create a set of n controllers, one for each goal (except the optimization goal). The controller C_i for goal r_i has the form

$$u_i(k) = u_i(k-1) - \frac{1-p_i}{\alpha_i}.e_i(k) \qquad (9.17)$$

The controller calculates the control signal $u_i(k)$ at time step k based on the previous value of the signal $u_i(k-1)$, the slope α_i of the system model for goal r_i, the pole p_i of the transfer function of the control system for this goal, and the error between the desired reference value and the measured value $e_i(k) = r_i(k) - y_i(k)$.

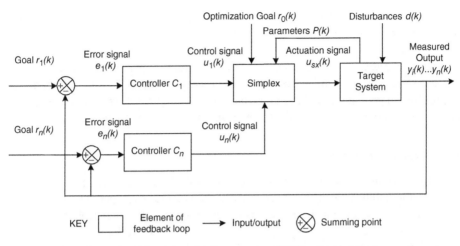

Figure 9.8 Operation phase of automatically generated MIMO control system.

The *operation* phase for the MIMO control system comprises two sub-phases: controlling and optimization, see Figure 9.8.

In the controlling sub-phase, the controllers effectively exercise control. Each controller C_i deals with one goal r_i, providing a control signal $u_i(k)$ to drive the corresponding output $y_i(k)$ to its goal while rejecting disturbances. The control signals produced by the controllers are fed into the Simplex block (see the optimization sub-phase below). To deal with non-linearities of the system model, the value of the slope α_i of the model M_i can be updated on-the-fly using a Kalman filter that is embedded in controller C_i, similarly to the basic approach for a single goal. Furthermore, a change point detection mechanism is added that can restart model building and controller construction if invasive changes of the control system occur.

The optimization sub-phase relies on the *Simplex* method, which is a classical method for solving the optimization problem of a linear program.

In general, the simplex method requires the linear program to be written in the standard form

$$max \ \{c^T x \mid Ax \le b; x \ge 0 \} \tag{9.18}$$

where $x = (x_1, \ldots, x_n)$ is the vector of variables that needs to be determined, $c = (c_1, \ldots, c_n)$ is a vector of known coefficients of the objective function, $b = (b_1, \ldots, b_p)$ is a vector of known non-negative constants, A is a $p * n$ matrix of known coefficients, and $(.)^T$ is the matrix transpose that flips a matrix over its diagonal. For our problem, we can use a simplified version of simplex with equalities, i.e. $Ax = b$. The simplex method solves the system of equations and finds an optimal allocation of values to the variables of x. In worst case, the algorithm runs in exponential time (order $O(2^n)$), but on average a solution is found in polynomial time (order $O(n^k)$). Since we are working with equations only, simplex will translate control signals to actuation signals without affecting the effect of the control signals on the output signals.

Translated to our control problem, we use simplex to find a proper actuation signal u_{sx} that corresponds to vector x. Each equation of the linear program except the last one represents a goal to be satisfied. The control signals $u_i(k)$ produced during the controlling phase correspond to the vector b, whereas a set of problem-specific system parameters $P(k)$ that are monitored at runtime correspond to matrix A and vector c^T. The last equation constrains the values assigned to the elements of vector x ensuring that a valid actuation signal is selected. Consequently, simplex takes the control signals produced by the controllers and the optimization goal and calculates an actuation signal u_{sx} that drives the target system toward an output that satisfies all adaptation goals, including the optimization goal.

9.3.2 Formal Guarantees

The controllers that deal with the goals (except the optimization goal) can provide formal guarantees for stability, steady-state error, settling time, and overshoot. The assessment is similar to the basic approach for automatic construction of SISO controllers, but now applied for each goal.

Stability of the control system is guaranteed if the pole p_i of the transfer function of each controller C_i belongs to the interval $[0, 1)$. To assess the settling time, we use the transfer function

$$y_i(k) = r_i(k) \cdot (1 - p_i^k) \tag{9.19}$$

When the output $y_i(k)$ reaches the goal $r_i(k)$ and the error is reduced to a value ϵ, we can rewrite the transfer function as

$$(1 - 0.01\epsilon) \cdot r_i(k) = r_i(k) \cdot (1 - p_i^k) \tag{9.20}$$

The time it takes for the output $y_i(k)$ to reach the goal $r_i(k)$ with an ϵ of 5% is

$$k_\epsilon = \frac{\log \ 0.01\epsilon}{\log | \ p_i \ |} \implies k_\epsilon = \frac{\log \ 0.05}{\log | \ p_i \ |} \tag{9.21}$$

Similar to the basic approach for automated construction of a SISO controller, the settling time and overshoot only depend on the value of the pole p_i of controller C_i.

Besides the guarantees provided by the controllers for the corresponding goals, simplex provides guarantees for the optimization goal. In particular, it has been proven that the simplex method always finds an optimal solution to a linear problem if one exists. In addition, the simplex method is able to detect infeasible solutions. Since the control system relies on equalities only, the solution of the linear program is always bounded; however, it may be infeasible to reach a goal. In that case, the controller will converge to the nearest achievable value for that goal. A simple detection mechanism can be added to the system that then alerts users if the goal is not reachable.

9.3.3 Example: Unmanned Underwater Vehicle

Consider an Unmanned Underwater Vehicle (UUV) that has to carry out a surveillance task in an ocean environment. In particular, the UUV is equipped with five on-board sensors that can measure an attribute, for example water current or salinity. Each sensor performs scans with a certain speed and accuracy, while consuming a certain amount of energy. A scan is performed every second.

The UUV environment is subject to disturbances, for instance sensor measurements may be inaccurate, the actual scanning speed of the UUV or its energy consumption may differ from the specification, or sensors may fail. Despite these uncertainties, the system is expected to satisfy the following requirements:

- A segment of surface over a distance of S (100 km) should be examined by the UUV within a given time t (10 hours in this scenario).
- To perform the mission, a given amount of energy E is available (5.4 MJ in the scenario).
- Subject to R1 and R2, the accuracy of measurements should be maximized.

We enhance the UUV system with an automatically constructed control system with two controllers, one that needs to decide which sensors to turn on and off during a mission and another one that needs to set the speed the UUV should use such that the goals are achieved. We assume that only one sensor is active at a time; however, a combination of sensors can be used sequentially during each adaptation period.

The UUV system is equipped with monitors that measure the scanning speed and energy consumption of the UUV, as well as the activity of sensors and the accuracy of their measurements at any point in time. Furthermore, the UUV offers actuators that allow the portion of time each sensor should be used during system operation and the speed of the UUV to be set.

Figure 9.9 shows experimental results of the MIMO controller realization.

We consider inaccurate sensor measurements, deviations of the scanning speed of the UUV and its energy consumption, and failing sensors, but we do not consider catastrophic events. Adaptation is applied every time step k, which corresponds to 100 measurements. The model building phase is marked by a gray shaded box (k from 0 to 25). Model building consists of two parts, each creating a model for a controller. For the model of the first controller, different levels of energy usage are systematically tested from minimum to maximum. For the model of the second controller, the scanning speed is systematically tested in the valid range. During model building the energy consumption, the amount of scanned surface, and the accuracy are measured.

After model building, the two controllers are created that will deal with requirements $R1$ and $R2$, and these, combined with simplex, maximize the accuracy of the measurements.

The following system of equations illustrates the linear program that simplex needs to solve for the UUV system

Maximize Accuracy

$$max[Acc_1 \times x_1 + Acc_2 \times x_2 + \ldots + Acc_5 \times x_5] \tag{9.22}$$

Subject to

$$\begin{cases} E_1 \times x_1 + E_2 \times x_2 + E_3 \times x_3 + E_4 \times x_4 = u_1 \\ V_1 \times x_1 + V_2 \times x_2 + V_3 \times x_3 + V_4 \times x_5 = u_2 \\ x_1 + x_2 + x_3 + x_4 = 1 \end{cases} \tag{9.23}$$

where x_j (with $j \in [1 \ldots 4]$) is the fraction of time sensor j should be used during system operation; Acc_j is the accuracy of sensor j; E_j is the energy consumed by sensor j; V_j is the scanning speed of sensor j (the values of Acc_j, E_j, and V_j are part of the specification of the sensors). u_1 and u_2 are control signals that simplex receives from the energy consumption controller and the scanning speed controller, respectively. The parameters $P(k)$ (see Figure 9.8) that need to be monitored in the UUV system are the energy consumption E_j

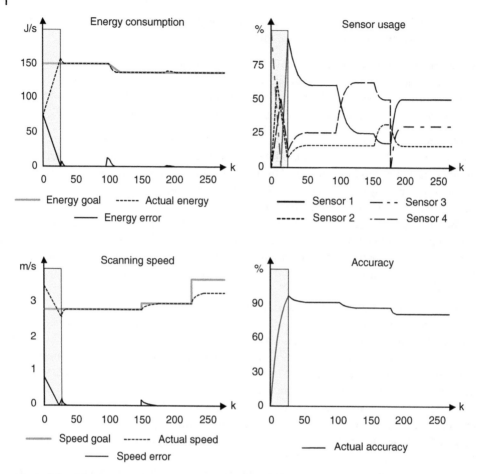

Figure 9.9 Experimental results of the MIMO controller realization.

and the scanning speed V_j of the active sensor j. Vector c^T of the linear program to be solved by simplex (see Equation 9.18) is replaced with the values of the accuracy Acc_j of the sensors. The last equation of the system of equations ensures that at any time instance during the mission, only one sensor is active and collecting data. Vector x represents the fraction of time each sensor should be used during system operation.

When the control system is enacted, the scanning speed is 3 m/s (meters per second), which ensures that the required surface will be scanned during the mission, *Sensor 1* is used for 60% of the time, *Sensor 2* for 15%, and *Sensor 4* for 25%. This results in an energy consumption of 150 J/s (Joules per second), close to the goal, and accuracy of around 90%. At $k = 100$, the available energy suddenly drops by 10%, changing the energy consumption goal from 150 J/s to 140 J/s. As a result, the fraction of time that *Sensor 1* is used is reduced to 25%, *Sensor 2* remains 15% active, and *Sensor 4* is now 60% active. The result is a small reduction in accuracy. At $k = 150$ the UUV is instructed to scan 10% more surface than initially planned. As a result, the scanning speed goal is increased to 3.2 m/s. The graph shows that the control system is stable and smoothly adapts. The fractions of time that the three active sensors are used are adapted as shown in the figure. At $k = 175$, *Sensor 4* fails. As a result, *Sensor 3* is activated and used for 30%, while the fraction that *Sensor 1* is used is increased to 50% and the fraction of *Sensor 2* is reduced to 20%. This causes another small drop in accuracy. Finally, at $k = 225$, an additional request is sent to the UUV to add another 20% of scanning surface. As a result, the scanning speed goal is increased to 4 m/s. However, the actual scanning speed saturates at 3.4 m/s, hence the user is notified about the infeasible goal.

9.4 Model Predictive Control

The two approaches for automated controller construction we have discussed so far, which are based on simple controllers of the form $u(k) = u(k-1) - \frac{1-p}{\alpha}.e(k)$, have been demonstrated to be effective for solving control problems in a variety of domains. However, these solutions have limitations: for instance, they cannot handle dependencies between goals or prioritize goals, and they may produce sub-optimal solutions. To tackle these limitations, the application of automated model predictive control (MPC) has been studied in the automated construction of controllers.

MPC is centered on the optimization of a utility function (or cost function) that accounts for the current operating point and all possible trajectories of the outcome of control decisions over a time horizon. MPC is in general considered to be particularly well suited to multi-objective problems with optimization because the controller takes into account all the inter-dependencies between actuators and goals simultaneously, resulting in a fully optimal solution.[7]

We present an approach to automatically construct a control system with a MPC controller that can handle multiple setpoint goals and one optimization goal. The approach is based on AMOCS-MA, short for Automated Multi-objective Control of Software with Multiple Actuators, a pioneering realization of automatic construction of a MPC controller for software adaptation.

9.4.1 Controller Construction and Operation

The realization of the MPC control system follows the three phases to automatically construct a control system: model building, control creation, and operation (see Figure 9.3).

To start model building, the designer needs to specify a set of v actuators $U = \{u_1, u_2, \ldots, u_v\}$ and a set of p goals $R = \{r_1, r_2, \ldots, r_p\}$ with $p \leq v$, which is a common requirement for MIMO control solutions. The value of goal r_i at time k is denoted $r_i(k)$, as goals may change during operation. Furthermore, the designer needs to specify a weight w_i for each goal r_i resulting in a vector of weights $W = \{w_1, w_2, \ldots, w_p\}$. These weights express the relative importance of goals, and are used to define a utility function determining how to relax goals in the event not all of them are feasible in a specific operational setting: higher importance goals will be considered more "costly" to relax. Optionally, the designer can specify weights for the actuators, $D = \{d_1, d_2, \ldots, d_v\}$, where a

---

7 For simplicity, we do not consider inaccuracies in the system model in this chapter.

lower weight indicates a preference for changing the corresponding actuator over actuators with higher weights. Finally, the designer needs to provide the vector of control signals U_i for each actuator, which will be used to test the system and identify the system model. This includes valid values for each control signal and a value for the sampling rate δ (see Equation 9.15). We use $U(\cdot)$ to denote the vector of values of the actuator signals, which vary over time.

Based on this data, the system is systematically tested, and a linear model S of the system is built (or identified) in the form of the following difference equation

$$S = \begin{cases} \tilde{x}(k+1) = A.\tilde{x}(k) + B.\Delta U(k) \\ Y(k) = C.\tilde{x}(k) \end{cases} \tag{9.24}$$

where $\tilde{x}(\cdot)$ represents the state variation of the system based on $\Delta x(k) = x(k) - x(k-1)$; $\Delta U(k)$ is the increment of the control signals of the actuators between time instances $k-1$ and k; and $Y(k)$ is the vector of system outputs at time k. A, B, and C are matrices with problem-specific coefficients that are obtained from running the experiment on the software system.[8]

In the next phase, *controller creation,* an MPC controller is automatically instantiated. The MPC controller uses the dynamic model of the system and an optimization cost function to calculate optimal control actions over a time horizon. The cost function $l(k)$ can be defined as follows

Minimize

$$l(k) = \sum_{i=1}^{L} \left(\sum_{j=1}^{p} w_j.[y_j(k+i) - r_j(k+i)]^2 + \sum_{l=1}^{v} d_l.[\Delta U_l(k+i-1)]^2 \right) \tag{9.25}$$

Subject to

$$\begin{cases} u_{min} \leq u(k+i-1) \leq u_{max} \\ U_{min} \leq U(k+i-1) \leq U_{max} \\ S \\ i = 1, \dots, L \end{cases} \tag{9.26}$$

where w_j is the weight associated with goal j; $y_j(\cdot)$ is the time varying output for goal $r_j(\cdot)$, which is computed according to S using the current measure as the initial value for S; d_l is the weight associated with actuator l; and $\Delta U_l(\cdot)$ is the time-varying increment of the control signals.

At any time instant k, the controller minimizes the loss function $l(k)$ over a time horizon defined by L, subject to the system model S and a set of constraints on the bounds of the actuation signals $u(k+i-1)$ and $U(k+i-1)$. The solution is an optimal plan of actuation

---

8 Notably, Equation 9.24 may not have a direct physical meaning for the system it models, nor is the size of the vector space required to match the number of control inputs. Equation 9.24 represents a linear state-space approximation in a possibly high-dimensional space of the relation between applied control signals and resulting measured outputs. Higher cardinalities of the state space model may provide suitable approximations for some non-linearities. However, the choice of appropriate system identification methods is beyond the scope of this chapter.

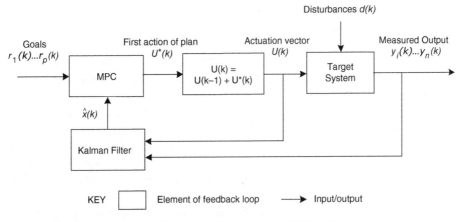

Figure 9.10 Operation phase of automatically constructed MPC control schema.

actions $U^*(k + i - 1)$ for the future with $i = 1, \ldots, L$. Typically only the first action of the plan $U^*(k)$ is applied, and the new control signal is then obtained as

$$U(k) = U(k - 1) + U^*(k) \tag{9.27}$$

The plan is recomputed every time new information becomes available. When one or more goals are infeasible (for example because there are conflicts between the goals), the controller will favor the goals with the higher weights.

The controller can be complemented with a mechanism for incremental model updating to deal with small variations of the target system and a change point detection mechanism that tracks abrupt changes, triggering model rebuilding.

The MPC control schema relies on knowledge of the varying state $x(\cdot)$ of the system. Since the state of the dynamic model usually has no direct relation to the components of the target system, it may not be possible to measure the state directly. One approach to solving this problem is by using a Kalman filter, which can compute an approximation of the state of the system $\hat{x}(k)$ using the dynamic model of the system, the values of actuation signals $U(\cdot)$, and the measurements of the system output $Y(k)$.

The resulting control schema that is enacted during the operation phase is shown in Figure 9.10.

By optimizing over a finite time horizon, MPC is able to make adaptation decisions in the current time instance, taking future time instances into account. As such, MPC is able to anticipate future events and perform control actions accordingly.

9.4.2 Formal Assessment

MPC belongs to the class of optimal controllers that make control decisions based on solving an optimization problem. Hence, the formal guarantees for the MPC control schema rely on the guarantees that are provided by the optimization algorithm that is used and the quality of the identified model, which is used to predict future evolutions of the system under different control decisions.

The automatically generated MPC solution relies on the cost function $l(k)$, which iteratively computes optimal adaptation plans with respect to system goals over a finite time horizon; from these plans the first action is then selected in each cycle to control the system. Hence, the prediction horizon keeps being shifted forward, which is called *receding horizon control*. Although there is no guarantee that this approach is optimal, practice shows that it gives very good results.

The MPC control schema ensures that all the goals $R(k)$ are reached if there exists a combination of actuator settings $U(k)$ within the given constraints that allows the system to reach its goals. If such solution does not exist, the optimization algorithm will find an actuator setting that minimizes the distance between the system output $Y(k)$ and the goals. How this distance is distributed over the goals depends on weights set in the cost function of the optimization problem.

Regarding the convergence time, the cost function l_k penalizes all the time instants where the output vector $Y(k)$ is not equal to the goal vector $R(k)$. Hence, the MPC control schema will ensure minimum settling time.

Since the MPC control schema needs to solve an optimization problem in each adaptation cycle, it is important to consider the time it requires to compute a solution. Fast solvers are one way to compute timely solutions. An easier solution is to reduce the time horizon, reducing the complexity of the optimization problem. Other techniques that can be applied are solving the optimization problem incrementally using so-called *interior point* algorithms, which iteratively update a feasible solution. The most recent intermediate solution can then be used when an adaptation action is required before the complete solution is found. Finally, a simple approach would be to store the last computed plan and apply the second control action if no new solution could be found by the time a new adaptation action is required. Clearly, while these alternatives allow faster response, they all result in potentially sub-optimal solutions.

9.4.3 Example: Video Compression

Consider a camera that is recording a video that needs to be streamed over a network and stored in an archive; an example is the recording of surveillance videos. The video is divided into frames, which are sent separately over the network. The frames can be compressed using a video encoder (the target system). Compressing the frames reduces the amount of disk space needed to store the video but also affects the video quality. We use the structural similarity index (SSIM) to quantify the similarity between the original and compressed frames. The value of SSIM ranges from 0 to 1: the higher the value, the greater the similarity. The system goals are:

- The quality of the compressed video should have a given value of SSIM (for instance 0.7). We refer to this goal as r_1 and set its weight $w_1 = 100$.
- The frame should have a given size (for instance 8 kilobytes). For this goal r_2, we set the weight $w_2 = 0.001$, which makes this goal slightly more important than video quality goal.

The video encoder provides the following actuators to exercise control over it:

- A parameter u_1 to set the compression density affecting the quality of the video. The parameter ranges from 1 to 100, where 1 produces the highest compression and 100 preserves all frame details. We use a weight for this actuator $d_1 = 1000$.
- A parameter u_2 to set the size of a sharpening filter to be applied on the frames. The size ranges from 0 to 5, where 0 indicates no sharpening. We set the weight d_2 of this actuator to 100 000.
- A parameter u_3 to set the size of a noise reduction filter, which also varies between 0 and 5. The weight d_3 is also set to 100 000.

To deal with these conflicting goals, we enhance the video encoder with an automatically constructed control system using an MPC. The controller should make a tradeoff between the goals to achieve the optimal value for the cost function.

Figure 9.11 shows experimental results of the automatic MPC controller realization.

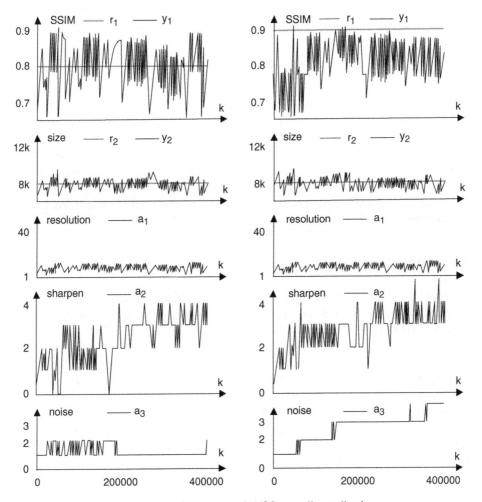

Figure 9.11 Experimental results of the automatic MPC controller realization.

The video compression scenarios used the Obama Victory Speech video,[9] using a prediction horizon $L = 4$. The graphs on the left hand side show the results for *SSIM* set to 0.8, while the graphs on the right hand side show the results for *SSIM* = *0.9*. For both scenarios, the frame *size* was set to 8*k*.

For the scenario with *SSIM* = *0.8* and *size* = *8k*, both goals are feasible. The graphs in the first and second row (on the left) show that the measured *SSIM* and *size* fluctuate around the respective goals. To that end, the controller sets the resolution around 5 and uses some sharpening and a small amount of noise reduction.

Increasing *SSIM* to 0.9 results in infeasible goals. The graphs in the first and second row (on the right) show that the *SSIM* goal cannot be achieved, while the *size* goal is still met. Increasing sharpening and adding noise reduction helps improve *SSIM*; however, the controller is not able to achieve the goal value of 0.9. This scenario shows that in the event of an infeasible goal, the controller will find the actuator settings that bring the output as close as possible to the goal.

Extensive measurements of the case study show an average overhead of 3 ms for the computation time with a standard deviation of 74 μs, which is very fast.

9.5 A Note on Control Guarantees

Similar to the traditional verification techniques that are applied to runtime models in order to make adaptation decisions (see Chapter 8), the guarantees for control properties apply to the model of the control system. Since these models are an abstraction of the control system, formal assessment needs to be complemented with validation of the real control system. Hence, approval for the acceptance of a control-based self-adaptive software system can only given when formal guarantees based on the model of the control system are combined with validation that provides evidence for the compliance of the real system with its requirements.

9.6 Summary

The need for a theoretical framework for self-adaptation on the one hand, and the complexity of applying formal verification techniques at runtime to tame uncertainty on the other hand led to the application of control theory to realize and analyze self-adaptive software systems.

Applying traditional control solutions to realizing self-adaptive software is complex due to the specific nature of software systems and the complexity of the mathematics needed for software engineers to apply control techniques. This led to the development of solutions that automatically construct a control solution.

A basic feedback control system consists of a target system that is manipulated by an externally added controller based on the difference between the measured output of the system and the required reference input, i.e. the control error.

---

9 https://www.youtube.com/watch?v=nv9NwKAjmt0

A central problem in using control theory to realize self-adaptive software is building a model of the target system. Such models can be specified manually based on knowledge of the system or identified based on experiments. A common approach to modeling the target system is by means of a set of difference equations.

A popular control schema is Proportional-Integral (PI) control. The control signal of a PI controller that manipulates the target system comprises a proportional term that changes the control signal directly and proportionally depending on the control error, and an integral term that takes into account the history of the errors over time. Tuning the contributions of these terms allows the designer to adjust the controller properties that determine the behavior of the control system in response to changes.

The main properties in control theory are stability, setting time, overshoot, and robustness. A software control system is stable if the output of the feedback control converges to an equilibrium, close to the reference input, after a change (during the transient phase). Settling time is the time it takes for the controller to converge to steady state. Overshoot refers to the maximum difference between the measured value and the goal during the transient phase. Robustness refers to the ability of the system to return to the steady state despite disturbances or inaccuracies in measurements or the system model.

Automatic controller construction consists of three phases: model building, controller creation, and operation.

For a SISO control system, model building automatically constructs a model of the target system by testing the system using a set of systematically sampled values of the control signal. By applying regression on the values of the measured output, a model of the target system is constructed. Incremental model updating is used to deal with relatively small variations of the target system. A Kalman filter is one mechanism that can be used to realize incremental model updating. The filter tracks the responses of the system and automatically adjusts the value of the slope of the target system model. Model rebuilding is used to deal with abrupt changes of the system. To that end, change point detection can be used, which identifies when an abrupt change occurs and then triggers a rebuild of the model of the target system.

The model of the target system is used to create a controller that keeps the output close to the reference input while rejecting disturbances during operation. The transfer function of the control system defines the relationship between the control output and the reference input. This function is characterized by a pole, which determines the dynamic behavior of the control system. To ensure stability, the value of the pole should be in the interval between 0 and 1. By selecting a concrete value for the pole within these boundaries, the designer can trade off setting time with overshoot.

Model construction for an automatically constructed MIMO control system (with multiple setpoint goals and one optimization goal) is similar to that for SISO systems; however, now one model for each goal is constructed, except for the optimization goal.

The different system models are then used to construct a controller, one for each goal. The controllers generate control signals based on the error between their respective measured outputs and the reference input. These control signals are fed into a simplex algorithm, which combines them with the optimization goal. Simplex solves the optimization problem of a linear program, resulting in a vector of actuation signals that are used to manipulate the target system.

Table 9.1 Key insights of Wave VI: Control-based Software Adaptation.

- Control theory provides a foundational framework for realizing different types of self-adaptive software systems.
- Control schemas for self-adaptive software can be automatically constructed.
- An automatically constructed linear model of the target system allows various software control problems to be solved.
- However, initial model building needs to be combined with online model updating and if necessary model rebuilding. This results in an adaptive control schema.
- An automatically constructed control schema that combines PI control with simplex allows multiple setpoint goals and an optimization goal to be handled.
- An automatically constructed control schema with MPC offers an optimal solution for self-adaptive software with multiple goals.
- The application of control allows formal guarantees to be provided for stability, settling time, overshoot, and robustness.
- Control guarantees apply to the model and need to be complemented with validation.

By selecting proper values of the poles of the controllers, the stability, settling time, and overshoot of the MIMO control system can be guaranteed. Simplex guarantees that an optimal solution is found if one exists. Otherwise, the controller will converge to the nearest achievable goals and notify the user that not all goals are achievable.

Automatically constructed MPC control applies optimization based on a cost function and takes into account the current operating point and the possible trajectories of the outcome of adaptation over a time horizon.

The cost function of the MPC control schema allows weights to be assigned to goals and actuators. Computing the cost function yields an optimal plan of actuation actions for the future, i.e. a series of vectors of control signals to be exercised on the target system. Typically, only the first action is applied, and the plan is then recomputed when new information becomes available that requires adaptation.

The formal guarantees of the MPC control schema depend on the guarantees provided by the optimization algorithm that is used. The schema guarantees that the goals will be reached if a solution exists; otherwise the distance between the output and reference input of non-achievable goals will be minimized.

Table 9.1 summarizes the key insights of Wave VI.

9.7 Exercises

9.1 Z-transforms for SISO controller: level H For the formal assessment of automatically constructed SISO controllers, we determined the closed loop transfer function as $G(z) = \frac{1-p}{z-p}$, with p the pole of the transfer function. For $z = 1$, the static gain of this transfer function is equal to 1. Based on this, we concluded that the inverse transform of $G(z)$ is $y(k) = r(k) \cdot (1 - p^k)$, which can be derived using Z-transforms. Demonstrate the correctness of this conclusion. Additional information can be found in [71]. For more information about Z-transforms, see for instance [94].

9.2 PI control for SAVE: level W

Setting. SAVE, short for Self-Adaptive Video Encoder, is an artifact for studying different adaptation strategies for control systems with multiple, possibly conflicting, goals by using multiple actuators. The video compressing example that we used to illustrate automated MPC is based on the artifact. SAVE simulates an adaptive video surveillance camera. The video produced by the camera has to be sent on a network. Since network bandwidth is potentially a scarce resource, the video encoder should achieve predictability in the amount of information streamed for every frame. The artifact comes with three pre-specified controller strategies (at increasing scale of sophistication): (i) *random* selects a random value for the actuators, (ii) *bangbang* implements a Bang-Bang adaptation strategy that bounces between the minimum and maximum value for the actuators, depending on the sign of the error for the objectives, and (iii) *mpc* is a model predictive controller that tries to minimize a cost function that considers the distance the system is from each of the objectives and the use of the actuation strategy.

Task. Download SAVE and install the artifact. Run the software using the random control strategy with the Obama Victory Speech video that is provided with the artifact and default settings of the goals. Repeat for the Bang-Bang control strategy. Generate results for the experiments with the two control strategies. Compare the results and discuss. Now run the software using the predefined model predictive controller. Set the *SSIM* goal to 0.6 and the *framesize* goal to 6000. Run the experiments with the Obama Victory Speech video and collect the results. Repeat for the *SSIM* goal set to 0.8 and the *framesize* goal set to 10 000. Compare the results and discuss. Now design a novel control strategy using PI control. Follow the instructions provided with the artifact to implement your solution. Test your controller and compare it with the pre-defined control solutions.

Additional material. The SAVE artifact can be downloaded via [136]. Further information about the artifact can be found in [135]. For information about Bang-Bang control, see [96]. The model predictive controller applied in the artifact is similar to the approach applied in [7]. For the basics of PI control, see for instance [8].

9.3 Discrete event controller synthesis for self-adaptation: level M

Setting. An alternative approach for control-based software adaptation is automatic synthesis of discrete event controllers. This approach supports a correct-by-construction controller derived from a set of requirements. Figure 9.12 gives a high-level overview of the approach.

The requirements model specifies the main activities of the target system and its environment, and the relationships between these activities, such as conditions and responses. An activity can be enabled or disabled, depending on the other activities and conditions. Executing an activity results in a change of the state of the requirements model.

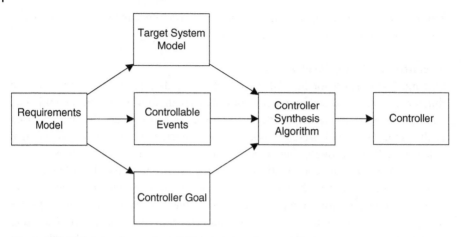

Figure 9.12 High-level overview of discrete event controller synthesis.

From the requirements model we can derive a model of the target system with relevant aspects of the environment, a set of controllable events, and a controller goal.

The target system model defines the assumptions the controller can rely upon to guarantee the system requirements, e.g. activities can only happen when they are enabled. Controllable events are events generated by the target system that can be regulated, i.e. the controller will be able to process them and reach a desirable state of the system. The controllable events are activity enabling and disabling events. Other events, such as the execution of an activity invoked in the environment, for instance by a human, can be monitored but not controlled. The controller goal encodes the constraints between activities that are expressed in the requirements model.

A controller synthesis algorithm requires these specifications in a proper formalism to enable automatic synthesis of a controller. The controller synthesis algorithm takes the target system model, the set of controllable events, and the controller goal and produces a controller. This controller will enable and disable activities of the target system in such a way that the system, as long as it only executes enabled activities, satisfies the requirements as described in the requirements model.

So far, discrete event controller synthesis has been applied to traditional control systems, for instance to synthesize controllers for the behavior of a robotic system and controllers that dynamically manage business processes.

Task. Study discrete event controller synthesis. Investigate how the approach can be applied to the adaptation of a managed software system and design a solution. Select an artifact and instantiate your solution. Evaluate your solution and assess.

Additional material. For an application of discrete event controller synthesis to a robotic system, see [33, 63]. For an application to manage business processes, see [146]. These approaches make use of tools that support the different steps of discrete event controller synthesis. For the selection of an artifact, see [196].

9.8   Bibliographic Notes

J. Hellerstein et al. wrote a pioneering reference book on the application of control theory to computing systems [94]. The book elaborates on model building, Z-transforms, and PID control. Examples of early applications of control theory used for the control of low-level resources of computing systems include: [152], which focused on the optimization of resource utilization in data centers; [128], which applied control theory to realize an elastic cloud computing environment; and [122], which dealt with power and performance management of virtualized computing environments.

For gentle introductions to control theory applied to computing systems, see [1, 126], and [109]. Y. Brun et al. studied feedback loop control applied to the adaptation of software [35]. S. Shevtsov et al. surveyed research on control-based software adaptation [178].

A. Filieri et al. presented an end-to-end control design process, from goal identification up to the validation of the controlled system [74]. The authors introduced a taxonomy of the main control strategies and their applicability to software adaptation for different types of goals.

The Push-Button Methodology that defined the foundation for the automatic construction of SISO controllers was introduced in [71]. For more information about the basics of Kalman filters, see [197]. S. Aminikhanghahi and D. Cook surveyed methods for time series change point detection [4]. For more background on formal assessment of control properties, see [94]. For more information about the ARPE tool, see [133]

SimCA, a control-based approach to the automatic construction of MIMO controllers, was introduced in [176]. SimCA leverages the Push-Button Methodology, PI control, and simplex. The Unmanned Underwater Vehicle example used for the evaluation of SimCA is based on [174] and [84].

M. Maggio et al. introduced AMOCS, which applies model-predictive control to software adaptation [134]. Model building in AMOCS leverages the principles of Push-Button Methodology. Levine's handbook of control elaborates on appropriate system identification methods [127]. For more information about the video compressing example, see the self-adaptive video encoder artifact [136].

K. Angelopoulos et al. presented a requirements-driven approach that applies MPC to software adaptation [7].

10

Wave VII: Learning from Experience

The seventh wave directs the focus to machine learning techniques as a means to enhance the realization of a self-adaptive software system. The fifth wave, Guarantees Under Uncertainties, which aimed at taming uncertainty through the use of formal techniques at runtime, was shown to be particularly challenging for systems with growing scale and increasing complex levels of uncertainty. A typical example is the analysis of the adaptation options for a system with very large adaptation spaces. Applying brute force formal verification may not work in such situations as verifying the complete adaptation space may not be feasible within the time window that is available to make adaptation decisions. Learning may then be used to reduce the adaptation space to the relevant adaptation options. The insights obtained from wave five were the driver for the seventh wave, which focuses on applying machine learning techniques at runtime to support various functions of a managing system that are required to make adaptation decisions.

The term "machine learning" was coined by Arthur Samuel in the late 1950s while he was working at IBM. Machine learning is usually considered a part of artificial intelligence. A definition for machine learning that is often used is: "A computer program is said to learn from experience E with respect to some class of tasks T and performance measure P if its performance at tasks in T, as measured by P, improves with experience E." The steady increase of computational power, in particular during the past decade, has made the practical application of powerful learning algorithms feasible for complex problem settings. As a result, machine learning is now widely used to deal with a variety of problems in the design and operation of software applications.

Technically, a machine learning algorithm builds a model from a data set that is collected from observing examples or tasks; this model is then used to make predictions on new, unseen examples or tasks. Since data sets are always finite and the future is uncertain, there are in general no guarantees on the performance of machine learning algorithms, but it may be possible to give probabilistic bounds.

Three important types of machine learning are supervised learning, unsupervised learning, and reinforcement learning. A supervised learning algorithm builds a model from a set of data that contains both the inputs and the desired outputs. Through iterative optimization of an objective function, a supervised learning algorithm learns a function that can be used to predict the output associated with new inputs. A classic supervised learning algorithm is Naive Bayes, which assigns class labels to problem instances, where the class labels are drawn from some data set. An unsupervised learning algorithm uses a set of data that

An Introduction to Self-Adaptive Systems: A Contemporary Software Engineering Perspective,
First Edition. Danny Weyns.

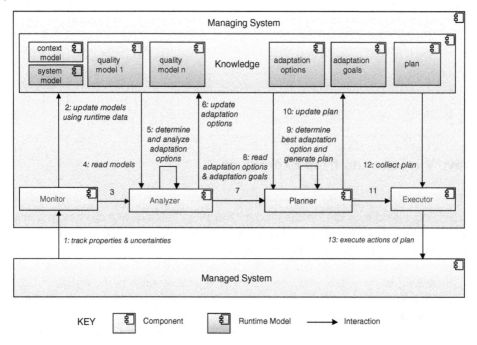

Figure 10.1 General overview of feedback loop functions that can be supported by learning.

contains only inputs. The algorithm searches for structure in the data, for instance finding clusters in the data. A typical example of unsupervised learning is a neural network that mimics the working of the brain to group unlabeled data according to similarities among the example inputs. In reinforcement learning, an agent perceives the state of the environment and chooses an action from the set of possible actions, moving the environment to a new state. The environment determines a reward associated with the transition, which the agent can observe together with the new state to choose the next action, etc. The goal of the agent is to maximize the cumulative reward. A classic reinforcement learning algorithm is Q-learning, which learns a policy that tells an agent what action to take under what circumstances.

Uncertainty is a fundamental characteristic of machine learning. In essence, a machine learning algorithm identifies patterns in imperfect or incomplete data of potentially large volume, while improving accuracy and efficiency over time as the algorithm gains experience. Based on the identified patterns, effective decisions can be made. Applied to self-adaptive software, machine learning can be used by a managing system to process data collected from sources that are subject to uncertainty and can support the decision making to adapt the system and achieve its goals.

The exploitation of machine learning techniques in self-adaptive systems is relatively new. Figure 10.1 shows a general overview of a MAPE feedback loop with the work flow of activities. Machine learning can be used to support a number of different activities of this work flow.

In this chapter, we explain three characteristic approaches to using machine learning in different activities of the MAPE work flow of self-adaptive software systems.

The first approach enhances the monitor function with a Bayesian estimator that keeps a runtime model up to date (in this approach, learning is centered on activity 2 in Figure 10.1). This simple approach demonstrates the power of machine learning when dealing with parametric uncertainties. The second approach enhances the analyzer function with a classifier that enables large adaptation option spaces to be reduced at runtime, improving the efficiency of the analysis phase of self-adaptation (here learning is centered on activity 5 in Figure 10.1). This approach shows how learning can help to deal with the complexity that comes with increasing scale of self-adaptive systems. Finally, the third approach enhances various feedback loop functions with a learning strategy that combines fuzzy control and fuzzy Q-learning to adjust and improve auto-scaling rules of a Cloud infrastructure at runtime (this learning approach spans several activities in Figure 10.1). This approach shows how machine learning can help to support decision making in self-adaptive systems that are subject to complex types of uncertainties.

LEARNING OUTCOMES

- To motivate the use of machine learning in self-adaptive systems.
- To explain and illustrate the use of a Bayesian estimator to keep runtime models up-to-date.
- To explain and illustrate the use of a classifier to reduce large adaptation spaces.
- To explain and illustrate the use of Q-learning to learn and improve scaling rules of a Cloud infrastructure.
- To analyze and apply a machine learning technique to solve a problem of a concrete self-adaptive system.

10.1 Keeping Runtime Models Up-to-Date Using Learning

Runtime models play a central role in self-adaptive systems. The feedback loop uses runtime models to keep track of the uncertain and changing operating conditions of the system, performs an analysis of these models to evaluate the options that are available for adaptation, and decides which option to select to adapt the managed system in order to achieve the adaptation goals.

The uncertainties that affect the system are usually encoded in the runtime models as model parameters. The values for these parameters can be determined before the system is deployed using different methods, for instance based on measurements in the field or experiments performed on the system, analysis of data obtained from similar systems, or data provided by domain experts. However, regardless of the method used, such values are estimates that may be incorrect. Moreover, given the dynamic operating conditions of self-adaptive software systems, the parameter values will change over time. Hence, to make effective adaptation decisions it is crucial that the uncertainty parameters and hence the runtime models are kept up-to-date. To that end, the central problem that needs to be solved is to ensure that the models *accurately* represent the actual conditions of the running system and the environment in which it operates.

In this section, we present a systematic approach to keeping runtime models up-to-date based on Bayesian estimation. A Bayesian estimator allows model parameters to be improved sequentially as new data arrives. In particular, Bayesian estimation uses knowledge that we already have about model parameters (commonly known as *the prior*) to calculate updates of the parameters based on new data (i.e. *the posterior* expectation). We focus on runtime models of quality properties that are formally specified, for instance as Markov models, queuing networks, or stochastic timed automata. Such models can be analyzed at runtime using formal verification techniques. The Bayesian approach we present is based on KAMI, short for Keep Alive Models with Implementations, a pioneering approach for keeping quality models up-to-date during operation.

10.1.1 Runtime Quality Model

The Bayesian approach that we study can be applied to different types of probabilistic models. We focus here on Discrete Time Markov Chains (DTMC), which in general can be represented as follows

$$DTCM = \begin{cases} S = \{1, \dots, k\} \\ S_{init} \in S \\ M : S \times S \rightarrow [0, 1] \\ L : S \rightarrow 2^{AP} \end{cases} \tag{10.1}$$

with S a finite set of states; M a transition probability matrix – its elements $m_{s,s'}$ represent the probability of moving from state s to s' with $\sum_{s' \in S} m_{s,s'} = 1$; and L a function that labels states with atomic propositions AP. We refer to the probability matrix with initial estimates as $M^{(0)} = \{m_{s,s'}^{(0)}\}_{s,s' \in S}$, which is called the prior transition probability matrix.

The Markov property means that the probability of choosing a transition from s to s' is independent of any earlier transitions. If we consider a path $X_1, \dots, X_n$ along a set of states, the likelihood function that this path is followed is defined as

$$P(X_1 = s_1, \dots, X_n = s_n) = m_{s_1,s_2} \times m_{s_2,s_3} \times \dots \times m_{s_{n-1},s_n} \tag{10.2}$$

Consider as an example the reliability model of the service-based health assistance system that we introduced in Chapter 1 as shown in Figure 10.2. This DTMC represents the work flow of the health assistance system with its discrete states $S = \{s_0, \dots, s_{14}\}$ and probabilities $m_{i,j}$ assigned to the branches between the states. The initial values of the probabilities are based on estimates, which may have been provided by domain experts based on information from the service providers.

As an example, the likelihood function applied to the path from s_0 via s_2 to s_{11} is

$$P(X_1 = s_0, X_2 = s_2, X_3 = s_6, X_4 = s_{11}) = m_{s_0,s_2} \times m_{s_2,s_6} \times m_{s_6,s_{11}}$$
$$= 0.3 \times 1.0 \times 0.04 = 0.012$$

meaning that from the initial state, the likelihood that the user will trigger the alarm and that the alarm service will then fail is 1.2%. Depending on the requirements at hand, this may or may not be an acceptable value. If the value violates one of the requirements, an alternative service configuration needs to be identified that complies with the requirements. The current service configuration is then replaced by the new configuration, which is then

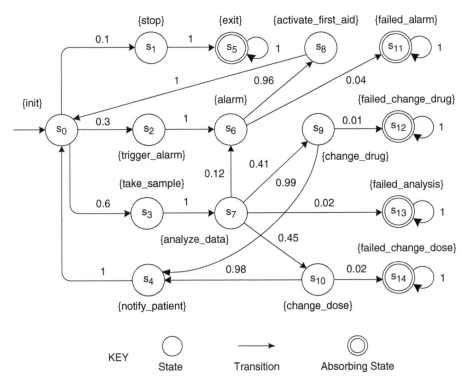

Figure 10.2 Reliability model of the service-based health assistance system as a DTMC.

selected for execution by the work flow engine. In practice, the managing system can use a model checker, such as PRISM, to automatically check the probabilities that different service failures occur in the model of Figure 10.2.

10.1.2 Overview of Bayesian Approach

The basic idea of the Bayesian approach is to track the system and environment elements that cause uncertainties, transform the collected data into model parameter values exploiting previous knowledge, and update the runtime model accordingly.

Figure 10.3 shows the relevant part of the feedback loop architecture (from Figure 10.1) that equips the monitor with a Bayesian estimator that in turn is used to keep a quality model up to date.

The architecture refines activity 2 in Figure 10.1 into three sub-activities (2.1 to 2.3). In particular, the approach distinguishes between updates of model parameters that do not require learning (activity 2.1), which typically refer to updates of non-stochastic properties, and updates of model parameters that capture uncertainties (activities 2.2 and 2.3). We focus here on the the latter and in particular on updates of a quality model represented as a DTMC.

Bayesian estimation exploits the prior transition probability matrix $M^{(0)}$ and statistical techniques to estimate the elements of the transition probability matrix M of a DTMC (called the posterior transition probability matrix) using new data. More specifically, the estimates $m_{i,j}^{(n_i)}$ are based on a set of data n_i of an event trace that is collected at runtime

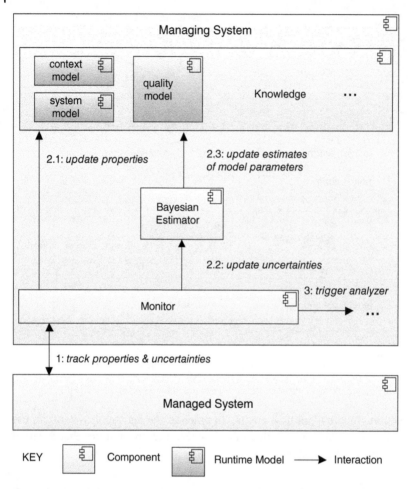

Figure 10.3 Monitor enhanced with Bayesian estimator to keep a quality model up to date.

about transitions from state i to state j and prior transitions $m_{i,j}^{(0)}$. Using Bayesian theory, the following updating rule can be derived to produce new estimates

$$
m_{i,j}^{(n_i)} = \frac{c_i^{(0)}}{c_i^{(0)} + n_i} \times m_{i,j}^{(0)} + \frac{n_i}{c_i^{(0)} + n_i} \times \frac{\sum_{h=1}^{d} n_{i,j}^{(h)}}{n_i}
\tag{10.3}
$$

where $c_i^{(0)}$ are smoothing factors that quantify the confidence in a priori knowledge versus new runtime data; d is the number of running instances of the system (i.e. the number of data traces); h is an instance of it; and $\frac{\sum_{h=1}^{d} n_{i,j}^{(h)}}{n_i}$ is the number of times state j is reached from state i for instances h. The first term in the updating rule relates to the initial estimates, while the second term relates to the runtime data.

As an example, consider a trace of 20 runtime data items $x_1, \dots, x_{20}$ in the health assistance system, each representing the invocation of an alarm. Suppose now that three of these invocations, say x_5, x_{15}, and x_{20}, resulted in failures of the alarm service. If we

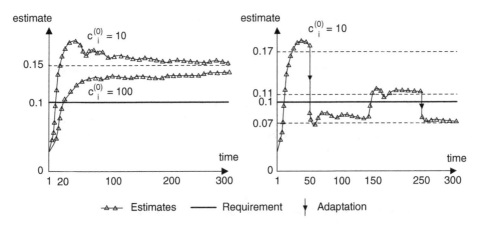

Figure 10.4 Simulated estimations of failure rate for link s_6 to s_{11} of the DTMC model. Left: a scenario without adaptation; right: a scenario with adaptation.

assume that c_i^0 is a constant of 10 in the DTCM of the health assistance system (shown in Figure 10.3), the new estimate for the probability of the transition from the *alarm* state (s_6) to the *failed_alarm* state (s_{11}) can be computed using the update rule as follows

$$m_{s_6,s_{11}} = \frac{10}{10+20} \times 0.04 + \frac{20}{10+20} \times \frac{3}{20} = \frac{0.4}{10+20} + \frac{3}{10+20} = \frac{3.4}{30} = 0.113 \quad (10.4)$$

The result shows that the initial probability 0.04 of the transition between *alarm* and *alarm_failed* was an underestimation and hence needs to be updated to 0.113. As a result, the probability of an alarm failure along the path from s_0 to s_{11} via s_2 (see Figure 10.2) is 0.3 × 1.0 × 0.113 = 0.034 (assuming that the probability of transition s_0 to s_2 did not change). If one of the system requirements is: "The percentage of alarm invocations directly invoked by the user that do not complete successfully is less than 3%," this will lead to a violation of the requirement. This in turn will require an adaptation of the system.

Since the updating rule of the Bayesian estimator takes into account prior and posterior probabilities, the probability of a transition that has changed will converge over time as more data is collected. Figure 10.4 shows the results of simulations of the estimated failure rate over the link from s_6 to s_{11} in the DTMC of the service assistance system shown in Figure 10.2. The horizontal axis shows the runtime data for the alarm invocations n_i. The vertical axis shows the estimated values $m_{s6,s11}$ over time. As indicated in the figure, the failure rate over the link in this scenario should be below 0.1. The simulation is based on runtime data of alarm invocations that follow a Bernoulli distribution with failure rate probability $p = 0.15$. Note that a Bernoulli distribution is a discrete probability distribution where the outcome is "success" with probability p and "failure" with probability $1 - p$.

The part of the figure on the left hand side shows that the prior value 0.04 gradually converges to the actual probability 0.15. The graphs show that the violation of the requirement for failure rate is detected within 20 runtime data points analyzed. Smaller values of the smoothing factor $c_i^{(0)}$ result in a shorter convergence process, but the process is less smooth, which may result in false positive requirement violations.

The graph on the right hand side of Figure 10.4 emulates how this problem can be resolved using adaptation (setting $c_i^{(0)} = 10$). In particular, around time step 50, the system selects a new service configuration, which reduces the estimated failure rate to approximately 0.08. At time step 150, a sudden change in the environment occurs, which increases the failure rate again above the threshold of 0.1. After converging, the system is adapted again around time step 250, bringing the failure rate below the threshold again, to approximately 0.07.

This scenario illustrates how Bayesian estimation supports the monitor component of a feedback loop by keeping runtime models updated under changing operating conditions over time. The up-to-date models can then be used by the other feedback loop components to adapt the system in order to achieve the adaptation goals.

10.2 Reducing Large Adaptation Spaces Using Learning

When analysis is required, the analyzer component needs to evaluate the possible options for adaptation, i.e. the adaptation space. Such analysis can be resource and time consuming, in particular when rigorous analysis methods are used, such as formal verification of runtime models. Hence, exhaustively analyzing all options within the available time window may be infeasible for systems with large adaptation spaces.

Machine learning can be used to tackle this problem. In particular, a traditional MAPE-K feedback loop can be enhanced with a learning module that selects relevant subsets of adaptation options from a large adaptation space to support the analyzer in performing efficient analysis. We focus here on an approach that relies on classification, which is a classic supervised learning technique that aims at assigning an observation (in our case an adaptation option) to a labeled class from a predetermined set of classes (here classes are defined by compliance with zero, one, or more adaptation goals). As such, classification can predict whether an adaptation option fulfills the adaptation goals. The adaptation options that comply with all the adaptation goals are labeled relevant by the classifier and will be analyzed by the verifier. From the adaptation options that are classified as not compliant with all the adaptation goals, a small fraction will be analyzed in order to explore new adaptation options that were classified as irrelevant before but may have become relevant due to changes in the uncertainties.

10.2.1 Illustration of the Problem

Consider the extended version of DeltaIoT as shown in Figure 10.5.

This IoT application, which we refer to as DeltaIoT.v2, comprises 37 nodes that are distributed across different floors of a building. The goal of the application is security monitoring of the building using different types of sensors, similar to DeltaIoT. The underlying technology of DeltaIoT.v2 is the same as that used in the basic DeltaIoT network. We consider the same uncertainties as for DeltaIoT – network interference and fluctuating traffic – and similar adaptation goals: over 12 hours, the average packet loss should be $\leq 10\%$, the average latency should be $\leq 5\%$ of the cycle time, and the average energy consumption should be minimized. The network of DeltaIoT.v2 is configured with communication cycles of 12 minutes: 9.5 minutes for sending sensor data to the gateway and further to the user

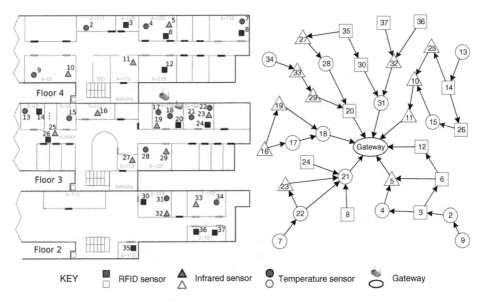

KEY ◼ / ▢ RFID sensor ▲ / △ Infrared sensor ● / ○ Temperature sensor 🖥 Gateway

Figure 10.5 Extended version of DeltaIoT with 37 nodes.

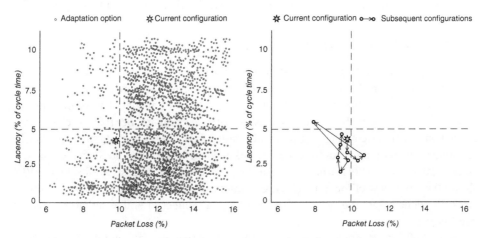

Figure 10.6 Left: adaptation space of DeltaIoT.v2 at some point in time. Right: changes of predicted values for the qualities of a single adaptation option over 10 cycles due to uncertainties.

application, and 2.5 minutes for sending adaptation messages from the gateway where the managing system is deployed to the motes.

We use the same management interface as for DeltaIoT, with an effector that allows the transmission power of the motes to be set from min 0 to max 15, while the distribution of the messages to multiple parents can be set in steps of 34% (0/100, 34/66, 66/34, 100/0). Based on the possible settings of the effectors, the total number of adaptation options for DeltaIoT.v2 is 4096 as illustrated on the graph at the left hand side of Figure 10.6.

This graph shows the complete adaptation space for DeltaIoT.v2 at some point in time. Each dot in the figure represents one adaptation option, i.e. a configuration of the system with a particular setting for the transmission power of each mote for each link and a setting of the percentage of messages that are sent over each link for motes with multiple parents. The position of each dot in the graph is based on the predicted values for two quality properties of that option (packet loss and latency), which are determined by verification.

Applying formal verification at runtime to analyze all the adaptation options within the available time window of the cycle time is not feasible. For instance, experiments have shown that verifying the complete adaptation space of over 4000 options with statistical model checking takes on average 12.5 minutes, which exceeds the cycle time. Note that the values for the quality properties shown in Figure 10.6 were determined offline.

In principle, the analyzer should only consider the "good" adaptation options, i.e. those with expected qualities that comply with the adaptation goals. In Figure 10.6, these are the adaptation options in the box defined by the thresholds of the two adaptation goals. However, note that the adaptation space is dynamic since the qualities of the adaptation options depend on the actual conditions of the network, i.e. the uncertainties due to interference on links and traffic load. The graph on the right hand side in Figure 10.6 shows how the predicted values of two quality properties for a single adaptation option of DeltaIoT.v2 changes over a period of 10 cycles due to the uncertainties.

Hence the problem for machine learning is to determine dynamically subsets of adaptation options from large dynamic adaptation spaces such that the analyzer can perform more efficient analysis without compromising the realization of the adaptation goals.

10.2.2   Overview of the Learning Approach

To support the analysis stage of adaptation, we equip the feedback loop with a classifier that predicts the relevant adaptation options by exploiting the available analysis results.

Standard classification utilizes batch learning to train its model. The trained model enables the mapping of observations to classes. Batch learning implies that the classifier is trained once with the complete collection of training data. Since a self-adaptive system will encounter uncertainties at runtime that are unknown beforehand, we need an incremental classifier that updates the learning model at runtime using new analysis results. Figure 10.7 shows the work flow of incremental classification, which consists of a number of offline activities (done before the system is running) and a number of online activities (done at runtime).

The first offline activity is collecting the data that is required for a number of analytical tasks that need to be performed to set up the learning module. In our context, we need data of verification results for the adaptation options over a number of cycles. These results will contain data of the verified qualities and data of the features. Features are variables of the system that affect the predicted qualities of adaptation options, which include the uncertainties (network interference and traffic load) and effector settings (power settings of motes to communicate packets over the links and the distribution factor per link, i.e. the percentage of packets sent by the source motes to each of its parents). The result of the first activity is a training set of data.

In the second activity, features are selected using a feature selection algorithm. Feature selection evaluates the effects that different features have on the actual prediction of the classifier. The effect of the features is expressed by giving each feature an importance score. Features with a very low score may be excluded. Excluding features helps simplify the model

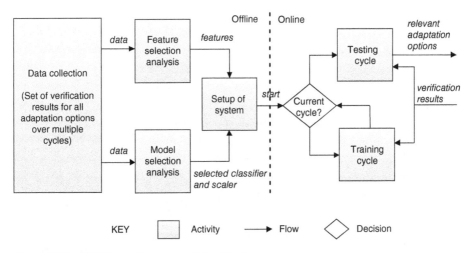

Figure 10.7 Work flow of incremental classification.

of the classifier and improves performance, but a potential disadvantage is that relevant features may be missed due to the limited coverage of the training set. The result of the second activity is a set of relevant features.

The third offline activity is model selection. A learning model (also referred to as a hypothesis) comprises the target function that we want to learn. The target function maps input to output, i.e. mapping adaptation options to classes that define the compliance of the adaptation options to the adaptation goals. To select a model, multiple combinations of classifiers and scalers are evaluated. Classifiers differ by variations in their loss and penalty functions. Whereas a loss function determines the inaccuracy of predictions, a penalty function determines the degree of impact that the loss will have on the model of the learner. A scaler transforms features to a specific scale, which is important for data that varies in magnitudes, units, and range. During model selection the designer evaluates different combinations of classifiers and scalers using quality metrics, such as, for instance, accuracy and F1-score. Accuracy determines the correctness of the predictions, while the F1-score combines recall (the percentage of samples that were retrieved by prediction from a specific class) and precision (the fraction of samples predicted to be of a specific class that are actually part of that class). The result of the third activity is a selected classifier and scaler for the learning module.

When the relevant features and the selected classifier and scaler are known, the machine learner module and corresponding models can be set up and integrated in the feedback loop system. This concludes the activities that are required before learning can be used at runtime.

Figure 10.8 shows the relevant part of the feedback loop architecture (from Figure 10.1), which integrates the classifier with the analyzer.

The architecture refines activity 5 in Figure 10.1 into six sub-activities (5.1 to 5.6). When the Analyzer is triggered, it reads the necessary runtime models and invokes the classifier to determine a relevant set of adaptation options (5.1). The classifier uses the classifier model to identify a subset of adaptation options that, based in its current knowledge, are

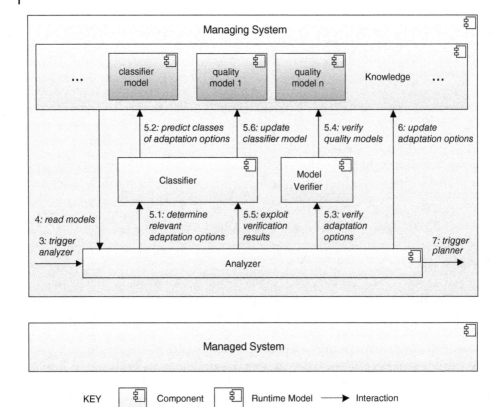

Figure 10.8 Analyzer enhanced with a classifier to determine relevant adaptation options.

expected to comply with the adaptation goals (5.2). In addition, a small sample of additional adaptation options are added to the relevant adaptation options for analysis in order to explore new adaptation options under changing operating conditions. The analyzer then invokes the model verifier to verify the subset of adaptation options (5.3). The verifier will determine the qualities of the set of adaptation options using the respective quality models (5.4). When verification is completed, the analyzer triggers the classifier to exploit the novel analysis results (5.5), continuing the learning process. To that end, the classifier will use the analysis results to update the classifier model (5.6). Finally, the analyzed adaptation options are updated, i.e. the analyzer associates the verification results with each of the selected adaptation options, and the planner is triggered to select the best adaptation option based on the adaptation goals and to generate a plan to adapt the managed system accordingly.

Note that the analyzer will start with a series of training cycles to train the model. In this phase, no adaptation space reduction is applied yet. The number of cycles that are required for training can be estimated during the offline stages of the learning work flow. After this bootstrap phase, the analyzer can fully exploit the learner to enhance the analysis process.

The graph on the left hand side of Figure 10.9 shows the adaptation options for one particular cycle that the classifier predicted to be of the class that complies with both the packet

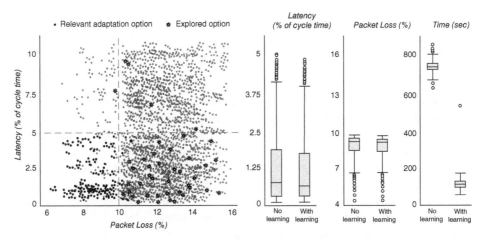

Figure 10.9 Left: relevant and explored adaptation options with classifier in one cycle. Right: comparison of quality properties and adaptation time without and with classification.

loss goal and the latency goal (marked with a dark dot). The graph also shows the sample of additional adaptation options of other classes that were selected for verification in this cycle (marked with a star).

Experiments with DelatIoT.v2 for a period of a week have shown that the combination of a stochastic gradient descent classifier with a standard scaler achieves an accuracy of around 86% (i.e. the percentage of correct predictions) with an F1-score around 0.85 (the F1-score combines recall and precision, with values in [0, 1], and a higher score means in general a better performing classifier). The three graphs on the right hand side of Figure 10.9 compare the results with and without learning, where statistical model checking was used for verification. It is important to note that the results for no learning were obtained via offline verification as exhaustive verification is not feasible within the available time slot of a cycle. The results show that there is no significant difference in the results for packet loss and latency. However, the time required for verifying the adaptation options when learning is applied is significantly lower compared to the case where no learning is used (mean 747 s for no learning and 174 s for classification). The outliers for the adaptation time with learning (approximately 570 s) represent the time required for training, when the complete cycle time is used to perform verification of randomly selected adaptation options. The experiments show that the time required for learning is less than 1% of the time required for verification.

10.3 Learning and Improving Scaling Rules of a Cloud Infrastructure

Meeting the performance requirements for users of a Cloud infrastructure and minimizing the operational costs for owners in the face of dynamic and irregular workloads requires dynamic acquisition and release of resources. A classic approach to tackling this problem is auto-scaling, which is typically based on threshold-based scaling rules (also called elasticity policies). A scaling rule relates performance metrics of the system (for instance workload and response time) to a particular resource assignment (for instance add or remove a resource). Cloud providers and platforms, such as Amazon EC2, Microsoft Azure, and OpenStack offer such auto-scaling facilities. The scaling rules used in such systems typically assume a linear dependency between performance metrics and resource assignments, which may not be sufficient to cover all possible conditions of the Cloud system in practice, for instance when hardware and software failures occur. Furthermore, defining scaling rules is difficult for application developers since they have insufficient knowledge of the infrastructure and the overall operating conditions, but also for Cloud providers as they have no clear view of the application requirements and behavior.

A solution to this problem is to equip the auto-scaler of the Cloud infrastructure with an online learning mechanism to create and improve scaling rules based on experience obtained at runtime. In this section, we study such an integrated mechanism that combines a fuzzy logic controller and fuzzy Q-Learning. Fuzzy techniques rely on the principles of fuzzy logic, which is a computing approach that allows specifying "degrees of truth" rather than the common "true" and "false" of standard Boolean logic. A fuzzy logic controller facilitates the expression of preferences at a higher level of abstraction, and Q-learning enables the controller to be adapted to improve its performance over time. The approach is based on FQL4KE, short for Fuzzy Q-Learning for Knowledge Evolution, a well-known self-adaptation approach for auto-scaling of Cloud infrastructures.

10.3.1 Overview of the Fuzzy Learning Approach

Figure 10.10 shows an overview of the architecture of the auto-scaling approach, which combines a fuzzy logic controller that realizes the analysis and planning functions in Figure 10.1 with a fuzzy Q-learner that learns scaling rules at runtime.

The self-adaptation problem of auto-scaling is to allocate computing resources to applications by monitoring the irregular and dynamic workload w and the performance of the system in terms of the average response time rt. To that end, a controller employs a set of scaling rules that determine how to dynamically increase or decrease the allocated resources in the form of scaling actions sa, in order to keep the performance rt at a desired level rt_{des}, while minimizing costs. Since defining scaling rules that satisfy the system goals under varying conditions is very hard, the feedback loop is enhanced with a learning module that dynamically learns and improves the scaling rules based on experience.

To illustrate the approach, we consider a scenario where a Cloud application consists of a number of "workers" that perform computations. To that end, workers require resources from the Cloud infrastructure in the form of virtual machines. We consider workloads with different patterns that represent different environmental conditions.

10.3.1.1 Fuzzy Logic Controller
The fuzzy logic controller takes as input the current workload, the average response time, a set of fuzzy membership functions, and a set of scaling rules, and produces as output a scaling action in the form of an increment or decrement of the number of deployed virtual machines.

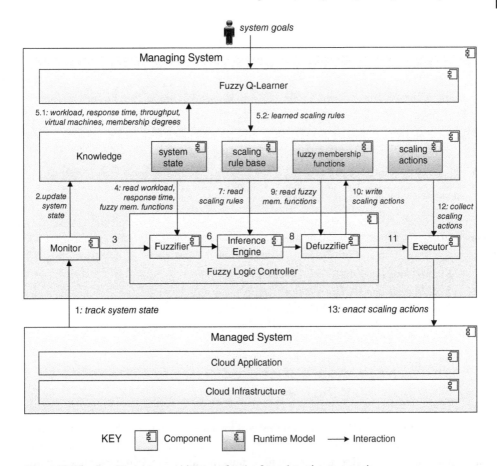

Figure 10.10 Feedback loop architecture for the fuzzy learning approach.

A fuzzy membership function defines fuzzy sets through membership functions. Each fuzzy set is characterized by a term such as "low," "medium," or "high." The membership function quantifies the degree of membership of an input signal to the fuzzy set. Figure 10.11 shows fuzzy membership functions for the two variables that are used in our example scenario.

Fuzzy membership functions can be defined based on stakeholder input. For instance, a workload below W_L is considered *Low,* between W_{ML} and W_{MH} is *Medium,* and above W_H is considered *High.* Between W_L and W_{ML} and between W_{MH} and W_H, the workload is to some degree low, medium, or high, as indicated by the membership functions. Similarly, a response time lower than RT_G is *Good,* while a response time of RT_{OK} is expected, and a above RT_B is *Bad.* Between RT_G and RT_B, the response time is to some degree good or bad.

Based on these membership functions, a set of in total nine scaling rules can be distinguished with combinations of memberships of the two variables, workload and response time. The scaling rules can be defined in the form of *if-then* statements as follows:

SR1: *IF (w is Low) AND (rt is Good) THEN (sa = -2)*

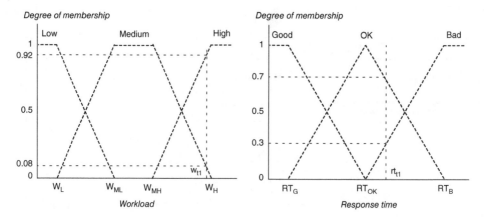

Figure 10.11 Fuzzy membership functions for the variables of the auto-scaling approach.

SR2: *IF (w is Low) AND (rt is OK) THEN (sa = −1)*
SR3: *IF (w is Low) AND (rt is Bad) THEN (sa = +1)*
SR4: *IF (w is Medium) AND (rt is Good) THEN (sa = −1)*
SR5: *IF (w is Medium) AND (rt is OK) THEN (sa = +0)*
SR6: *IF (w is Medium) AND (rt is Bad) THEN (sa = +1)*
SR7: *IF (w is High) AND (rt is Good) THEN (sa = +0)*
SR8: *IF (w is High) AND (rt is OK) THEN (sa = +1)*
SR9: *IF (w is High) AND (rt is Bad) THEN (sa = +2)*

The first scaling rule *SR1* states that if the workload is low and the response time is good, the resources (virtual machines) can be reduced by two entities. The other rules can be interpreted similarly. Scaling actions can have any arbitrary value; in our example scenario we consider only a simple set of five possible actions: {−2, −1, 0, +1, +2}. The scaling rules are usually provided by stakeholders of the system before the system is deployed. One of the key benefits of using fuzzy scaling rules is that a limited number of rules are required to specify a scaling policy, which reduces the state space of the system.

To make adaptation decisions, the fuzzy logic controller follows a three step process, corresponding to the three main components shown in Figure 10.10. In step 1, the fuzzifier performs fuzzification of the inputs, i.e. the concrete values of the data are transferred to fuzzy memberships using the membership functions. For instance, consider in Figure 10.11 a workload at time *t1* with value w_{t1}. This value is transferred to a fuzzy membership *Medium* with a degree of 0.08 and a membership *High* with degree of 0.92. Similarly, the response time at time *t1* with value rt_{t1} is transferred to a fuzzy membership *OK* with degree of membership 0.7 and a membership *Bad* with degree of 0.3.

In step 2, the interference engine takes the output of the fuzzifier (fuzzy memberships with degrees of membership) and the scaling rules, and derives fuzzy actions. Concretely, the inference engine calculates fuzzy actions as a weighted average as follows

$$a = \sum_{i=1}^{N} a_i(s) \times a_i \qquad (10.5)$$

Table 10.1 Firing levels of the rules at time *t1* in the auto-scaling scenario.

Rule number	SR1	SR2	SR3	SR4	SR5	SR6	SR7	SR8	SR9
Firing level	0	0	0	0	0.056	0.024	0	0.644	0.276

with N the number of scaling rules, $a_i(s)$ the firing level of rule i for input signal s (i.e. the state of the relevant variables) and a_i the required effect of rule i, also called the consequent of the rule. The fuzzy engine determines the firing level of each rule based on the fuzzy memberships and the input signal.

Consider again as example a workload w_{t1} and response time rt_{t1} as illustrated in Figure 10.11. Assume that the inference engine had determined the firing level of the rules as shown in Table 10.1.

In this situation, scaling rules *SR5, SR6, SR8,* and *SR9* apply. The inference engine will then calculate the following scaling action:

$$a_{t1} = \alpha_{SR5}(s_{t1}) \times a_{SR5} + \alpha_{SR6}(s_{t1}) \times a_{SR5} + \alpha_{SR8}(s_{t1}) \times a_{SR8} + \alpha_{SR9}(s_{t1}) \times a_{SR9}$$
$$a_{t1} = 0.056 \times 0 + 0.024 \times 1 + 0.644 \times 1 + 0.276 \times 2 = 1.22$$
(10.6)

As an example, rule *SR9* states that for a high workload and a response time that is bad, two additional resources needs to be added. With a firing level of *SR9* for w_{t1} equal to 0.276, this contributes $0.276 \times 2 = 0.552$ resources to be added to scaling action a_{t1}. The other summands can be interpreted in a similar way. The result of the calculation states that a scaling action a_{t1} to be applied is adding 1.22 resources to the system.

In step 3, the defuzzifier transfers the fuzzy output of the interference engine to effective scaling actions that will be enacted by the executor component. In our setting, the output is rounded to the nearest integer to match the discrete nature of scaling actions. For the example, this means that one resource will be added to the system.

10.3.1.2 Fuzzy Q-learning

A standard fuzzy logic controller assumes that the scaling rules are provided by stakeholders. However, as we explained, defining a set of rules that ensures robust auto-scaling actions under changing operating conditions can be very difficult in practice. Fuzzy Q-learning, which is a particular form of reinforcement learning, allows scaling rules to be learned at runtime, possibly, but not necessarily, from an initial set of rules defined by stakeholders of the system.

The overall goal of reinforcement learning is to ensure that the controller acquires and modifies knowledge about which actions to select in different situations based on how good the actions have been in the past. The "goodness" of actions is measured by means of a reward (i.e. a reinforcement) that can be observed from the environment after an action is taken. Here we consider long-term cumulative rewards, i.e. the controller takes scaling actions based on the expected cumulative reward that can be received by taking actions in the evolving state over time. With fuzzy Q-learning, the system learns to select actions not based on deterministic rules where only one rule fires to select an action, but where all rules have some contribution based on their firing level. Algorithm 1 shows a high level specification of the fuzzy Q-learning algorithm.

Algorithm 1 Fuzzy Q-Learning

1: **Given:** γ, η, ϵ
2: γ is a discount factor, η a learning rate, and ϵ an exploitation/exploration strategy
3: **Initialize the q-values:** $q[i, j] = 0, i = [1 \ldots N], j = [1 \ldots J]$ for N rules and J actions
4: **Select action a_i for each firing rule:**
5: $a_i = argmax_k\, q[i, k]$ with probability ϵ
6: $a_i = random\, \{a_k, k = 1 \ldots J\}$ with probability $1 - \epsilon$
7: **Calculate the control action by the fuzzy controller**
8: $a = \sum_{i=1}^{N} \alpha_i(s_t) \times a_i$ where a is the selected control action, s_t the state at time t, and
9: $\alpha_i(s_t)$ the firing level of rule i in state s_t
10: **Approximate the Q-function for firing rules with the selected action:**
11: $Q(s_t, a) = \sum_{i=1}^{N} \alpha_i(s_t) \times q[i, a]$ where $eq[i, a]$ is the current q-value for action a in rule i
12: **Take action a after which the system goes to state s_{t+1}**
13: **Observe the reward signal r_{t+1}**
14: **Calculate the Q-function for the new state denoted as $V(s_{t+1})$:**
15: $V(s_{t+1}) = \sum_{i=1}^{N} \alpha_i(s_{t+1}).max_k(q[i, q_i])$
16: **Calculate the error signal:**
17: $\Delta Q = r_{t+1} + \gamma \times V(s_{t+1}) - Q(s_t, a)$
18: **Update the Q-table values:**
19: $q[i, a_i] = q[i, a_i] + \eta \times \Delta Q \times \alpha_i(s_t)$
20: **Repeat for the next state until convergence.**

A central artifact of Q-learning is the Q-table. In our scenario, the Q-table consists of rows that represent scaling rules and columns that represent scaling actions. For each entry of the Q-table, which corresponds to a specific scaling rule and one possible scaling action, the learning algorithm learns a q-value. This value expresses how good the action is when that particular rule applies.

At a course-gained level, after the initialization of the q-values (line 3), the algorithm consists of 4 steps sequentially performed in a loop (from line 4 to line 20): choose an action based on the current q-values (lines 4 to 11), perform the action (line 12), observe the reward (line 13), and update the q-values and return (lines 14 to 20). We explain the algorithm applied to the auto-scaling scenario.

Lines 1–2. The discount factor γ determines the relative importance of future rewards. In the example scenario, we assume that the discount factor is set to $\gamma = 0.8$. The learning rate η determines the extent to which new information overrides old information, i.e. the learning rate determines how fast estimates are modified. We assume here that $\eta = 0.1$. The exploitation/exploration strategy ϵ determines the tradeoff between exploration, i.e. take a random decision, where more information is gathered (via the reward) that may improve decisions in the future and exploitation, where decisions are made based on the current information. We apply and compare different values for the exploitation/exploration strategy ϵ.

Line 3. The q-values that are maintained in a Q-table determine how good actions are in different states based on past experiences. Each q-value provides the expected future reward

Table 10.2 Excerpt of Q-table at time *t1* in the auto-scaling scenario.

System state	Scaling actions				
rule nb: [w, rt]	−2	−1	0	+1	+2
...	...	...	...	...	...
5: [Medium, OK]	0	0	0	0	0
6: [Medium, Bad]	−0.039	0.011	0	0	0
...	...	...	...	...	...
8: [High, OK]	0	0.001	0	−0.109	0
9: [High, Bad]	0	0	0.968	0.819	2.109

for an action in a given state. A high reward indicates that the action will likely have a good effect, a reward close to zero implies that the action is not effective, while a negative reward indicates that taking the action may result in a negative effect. In our scenario, the state space is finite and consists of only nine states (i.e. the combination of 3 x 3 memberships for fuzzy variables w and rt) and the controller can select scaling actions among five possibilities (remove one or two resources, do not change the number of active resources, or add one or two resources). Hence, the Q-table in the example scenario is a matrix of 9 x 5. The entries in the Q-table can be initialized with zeros, or alternatively filled with initial values that are obtained from stakeholders before deployment.

Lines 4–9. For each firing rule, the controller selects an action based on an exploitation/exploration strategy. Either the controller selects from the Q-table the action with the highest q-value for the given state with a probability ϵ, or it selects a random action with a probability $1 - \epsilon$. The three-step procedure that the fuzzy controller uses to select a control action is explained above.

Lines 10–11. Based on the firing level of the rules (for the current state of the system), the selected control action, and the current values in the Q-table, an approximation of the q-value is calculated. Consider Table 10.2, which shows an excerpt of the Q-table at time *t1* in our scenario.

Recall that at time *t1*, the firing levels determined by the fuzzy engine are 0.056 for *SR5*, 0.024 for *SR6*, 0.644 for *SR8*, and 0.276 for *SR9*. Further, the selected scaling action is +1. The approximated q-value in this scenario is then calculated as follows:

$$Q(s_{t1}, +1) = \alpha_{SR5}(s_{t1}) \times q[SR5, +1] + \alpha_{SR6}(s_{t1}) \times q[SR6, +1] +$$
$$\alpha_{SR8}(s_{t1}) \times q[SR8, +1] + \alpha_{SR9}(s_{t1}) \times q[SR9, +1]$$
$$Q(s_{t1}, +1) = 0.056 \times 0 + 0.024 \times 0 + 0.644 \times -0.109 + 0.276 \times 0.819 = 0.156$$

$$(10.7)$$

Lines 12–13. Next, the selected scaling action is taken. After the system is adapted (virtual machines are activated or deactivated), the reward signal can be observed. For the example scenario, we determine the reward signal $r(t)$ based on the change in utility of the system

Table 10.3 Sample data for fuzzy Q-learning scenario.

Time	Workload (%)	Throughput (%)	Number of workers	Response time (ms)
t1	86.3	19.1	1	670
t2	60.1	45.1	2	707

before and after the scaling action, which is calculated as follows

$$r(t) = U(t) - U(t-1) \tag{10.8}$$

where $U(t)$ is the utility value of the system at time t, which expresses the "goodness" of the system; it is defined as follows:

$$U_t = w_1 \cdot \frac{th_t}{th_{max}} + w_2 \cdot \left(1 - \frac{vm_t}{vm_{max}}\right) + w_3 \cdot (1 - H_t)$$

$$H_t = \begin{cases} \dfrac{rt_t - rt_{des}}{rt_{des}} \ for \ rt_{des} \le rt_t \le 2.rt_{des} \\ 1 \ for \ rt_t \ge 2.rt_{des} \\ 0 \ for \ rt_t \le rt_{des} \end{cases} \tag{10.9}$$

where th_t is the throughput at time t, vm_t is the number of workers (corresponding to the number of virtual machines used), and rt_t is the response time. To aggregate these elements, their values are normalized and multiplied by weights w_1, w_2 and w_3 respectively that determine the relative importance of each element in the utility function.

Table 10.3 shows sample data at time stamps *t1* and *t2* of the experiment.

With the maximum number of virtual machines vm_{max} set to 7, the desired response time rt_{des} set to 600 ms, and the weights w_1, w_2 and w_3 all set to 1, the utility of the system at time *t1* is calculated as follows

$$U(t1) = 1.\frac{19.1}{100} + 1.\left(1 - \frac{1}{7}\right) + 1.\left(1 - \frac{670 - 600}{600}\right) = 0.191 + 0.857 + 0.883 = 1.931 \tag{10.10}$$

After adding a worker (virtual machine) at time *t1*, the new state at *t2* is observed resulting in a utility

$$U(t2) = 1.\frac{45.1}{100} + 1.\left(1 - \frac{2}{7}\right) + 1.\left(1 - \frac{707 - 600}{600}\right) = 0{,}451 + 0.714 + 0.822 = 1.987 \tag{10.11}$$

Hence, the reward signal is: $r(t2) = 1.987 - 1.931 = 0.056$. Thus, the scaling action at time *t1* to add an additional virtual machine has a slightly positive effect on the utility of the system. In particular, the improvement of the throughput of the system is more significant than the slight increase in the response time.

Lines 14–17. The q-value for the new state of the system at time *t2* is now calculated based on the firing level of the rules for the new state and the current values in the Q-table. After fuzzification, the observed state of the system (see Table 10.3) is now a workload with

Table 10.4 Firing levels of the rules at time $t2$ in the auto-scaling scenario.

Rule number	SR1	SR2	SR3	SR4	SR5	SR6	SR7	SR8	SR9
Firing level	0	0	0	0	0.200	0.600	0	0.050	0.150

membership Medium and degree 0.8 and High with degree 0.2; response time has now a membership OK with degree 0.25 and membership Bad with degree 0.75. As a result, we assume that the inference engine computed new levels for the firing rules as shown in Table 10.4.

In this situation again the scaling rules *SR5, SR6, SR8,* and *SR9* apply. The q-value for the new state of the system is then calculated as follows:

$$V(s_{t2}) = \alpha_{SR5}(s_{t2}) \times max_k(q[SR5, q_{SR5}) + \alpha_{SR6}(s_{t2}) \times max_k(q[SR6, q_{SR6})$$
$$\alpha_{SR8}(s_{t2}) \times max_k(q[SR8, q_{SR8}) + \alpha_{SR9}(s_{t2}) \times max_k(q[SR9, q_{SR9})$$
$$V(s_{t2}) = 0.200 \times 0 + 0.600 \times 0.011 + 0.050 \times 0.001 + 0.150 \times 2.109 = 0.323$$

(10.12)

where α_{SR8} is the firing level of rule *SR8* and $max_k(q[SR8, q_{SR8}])$ is the maximum q-value in the row of rule *SR8* of the Q-table (and similarly for the other rules).

We can now calculate the error signal:

$$\Delta Q = r_{t2} + \gamma \times V(s_{t2}) - Q(s_{t1}, +1)$$
$$\Delta Q = 0.056 + 0.8 \times 0.323 - 0.156 = 0.158$$

(10.13)

Lines 18–19. Finally the Q-table entries are updated. We illustrate the update for the entry of SR9 (*[High, Bad]*) and scaling action +1:

$$q[SR9, +1] = q[SR9, +1] + \eta \times \Delta Q \times \alpha_{SR9}(s_{t2})$$
$$q[SR9, +1] = 0.819 + 0.1 \times 0.158 \times 0.75 = 0.831$$

(10.14)

The value for this entry of SR9 has increased as the scaling action turned out to be useful. The other entries are updated in a similar way. Figure 10.12 shows two examples of the evolution of Q-values over time.

10.3.1.3 Experiments

Extensive experiments have shown that fuzzy Q-learning has a positive impact on the efficiency of resource allocation in a Cloud environment. We illustrate a number of results reported for a prototype implemented on Microsoft Azure and OpenStack. In this setting, workers calculate Fibonacci numbers with workload patterns as shown in Figure 10.13.

The x-axis shows the experimental time over a period of 24 h and the y-axis shows the number of user requests, which represents the workload intensity. The slowly varying pattern gradually increases the workload until it reaches a maximum and then gradually decreases the load. With the large variations pattern, on the other hand, the workload continuously fluctuates, often with sharp peaks. These patterns pose different challenges to the fuzzy Q-learner.

Figure 10.14 shows two different exploration/exploitation strategies that are considered in the experiments. These strategies determine how the the controller selects actions (see lines 4–6 in Algorithm 1).

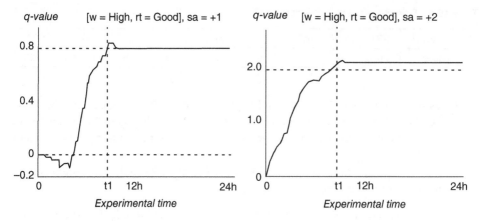

Figure 10.12 Excerpts of the evolution of q-values.

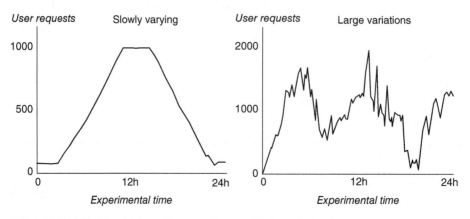

Figure 10.13 Workload patterns for experiments with fuzzy Q-learning.

Strategy S1 starts with maximum exploration rate, $\epsilon = 1$, to ensure the learner sufficiently explores actions over a substantial period of time. While the learning process converges, the exploration rate is gradually decreased. When the rate reaches a value $\epsilon = 0.2$, it stabilizes, ensuring that sufficient exploration remains to deal with new operating conditions. Strategy S2, on the other hand, maintains a constant relatively high exploration rate with a value of $\epsilon = 0.5$ in order to handle unpredictable environments at all times.

Table 10.5 shows experimental results of different performance metrics for strategies S1 and S2 for the two workload patterns. In addition, the results are compared with a setting where the Azure auto-scaler is used without learning.

The metrics considered are the average response time of 95 % of the measurements ($rt_{95\%}$), the average number of virtual machines acquired during the experiment (vm_{av}), the sum of the changes required in the resources *(scaling actions)*, and the time required to convergence to an optimal policy (*convergence* in % of the experimental time).

The results show that strategy S1 with a gradually decreasing exploration factor outperforms strategy S2 with a constant relatively high exploration factor on all metrics. Both S1 and S2 outperform the Azure scaler on response time and average number of virtual

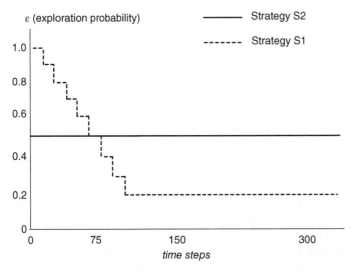

Figure 10.14 Fuzzy membership functions used in the experiments with the auto-scaling approach.

Table 10.5 Performance results of fuzzy Q-learning.

Strategy	Metric	Slowly varying	Large variations
	$rt_{95\%}$, vm_{av}	512 ms, 3.6	991 ms, 4.3
S1	node changes	355	420
	convergence	24	40
	$rt_{95\%}$, vm_{av}	507 ms, 3.7	1203 ms, 4.3
S2	node changes	372	432
	convergence	40	68
	$rt_{95\%}$, vm_{av}	1101 ms, 3.7	1341 ms, 5.5
Azure	node changes	287	367
	convergence	N/A	N/A

machines. The Azure scaler without learning shows better results for the number of change actions, but this approach does not improve over time, while the learning approaches converge to an optimal policy after some time. Experimental results show that the runtime overhead of learning is negligible given the delays to actuate and deactivate virtual machines (which is on the orders of minutes).

10.4 Summary

The need for the application of self-adaptation in large-scale systems that may be subject to complex types of uncertainties has motivated the use of machine learning in self-adaptive systems.

Uncertainty is fundamental to machine learning as learning algorithms in essence identify patterns in imperfect and potentially large volumes of data, paving the way to support effective decision making in self-adaptive systems. Machine learning can be used to support different activities in the work flow of a MAPE feedback loop.

A common approach to encoding uncertainty in the knowledge of a managing system is through the parameters of probabilistic models. While the values of these parameters can be estimated before the system is deployed, for instance based on stakeholder input, the estimates will be invalidated when the operating conditions of the system dynamically change.

Bayesian estimation supports the monitoring function of a MAPE feedback loop by continuously improving the model parameters of a probabilistic model during operation. A Bayesian estimator exploits prior knowledge and statistical techniques to estimate the uncertainty parameters of a probabilistic model using traces of new data. Changes in the parameters will converge over time as new data arrives. The up-to-date models can then be used by the analyzer to detect goal violations and activate adaptations.

Exhaustively verifying all the adaptation options of a large-scale system within time and resource constraints may be hard or even infeasible. This problem applies in particular to self-adaptive systems that use rigorous analysis methods.

Classification, a classic supervised learning technique, allows the prediction of whether an adaptation option fulfills the adaptation goals. Analysis will then be performed on the options that are predicted to comply with the goals. In addition, a small fraction of the options that do not comply with all the goals are analyzed as well in order to explore new adaptation options. The results of the analysis can be fed back to the classifier to continue the learning process over time.

Ensuring service level agreements with clients of a Cloud infrastructure while being cost effective requires dynamic acquisition and release of resources based on the changing workload on the system. A classic approach to realizing this is auto-scaling. Auto-scaling is based on rules that relate performance metrics of the system to resource assignments. Yet, defining such rules is complex in practice.

Enhancing the auto-scaler of a Cloud infrastructure with an online learning mechanism enables scaling rules to be created and improved based on experiences obtained at runtime. A particularly interesting approach that instantiates this idea combines a fuzzy logic controller with fuzzy Q-learning.

A fuzzy logic controller takes as input a set of scaling rules and the current values of the properties of interest (for instance workload and response time). These values are transferred into membership degrees using a set of fuzzy membership functions (for instance, workload high with a degree of 0.7 and medium with a degree of 0.3). These degrees determine the fraction each membership contributes when calculating scaling actions for the different scaling rules.

Fuzzy Q-learning, which is a particular type of reinforcement learning, allows the scaling rules to be learned at runtime. Q-learning learns the entries of a Q-table, which in the case of auto-scaling assigns a q-value to each scaling rule and each possible action. After initializing the Q-table, the learning algorithm runs in a loop comprising four steps: (i) the controller selects a scaling action; (ii) the action is enacted on the system; (iii) the reward of the action is computed based on the observed utility (e.g. based on observed throughput and response

Table 10.6   Key insights of Wave VII: Learning from Experience.

• Machine learning enables large amounts of data and complex levels of uncertainty to be handled. Learning can be used to support different functions of a feedback loop.

• Bayesian estimation can be used to support the monitoring function in keeping parameterized probabilistic models up-to-date.

• Classification can be used to support the analysis function in reducing the adaptation space to only the relevant adaptation options.

• Fuzzy Q-learning can be used to support the auto-scaler of a Cloud infrastructure to learn and improve scaling rules over time.

• Learning is increasingly being applied in self-adaptation. Different learning techniques have shown good performance in practice, while generating relatively low overhead.

• Learning takes time. Hence, the time to learn should only be a fraction of the time between the changes in the system that require learning new adaptation behavior.

---

time); and (iv) based on the expected and obtained results, the q-values are updated. These updated values, which converge over time, enable the controller to systematically improve its performance.

Table 10.6 summarizes the key insights of Wave VII.

10.5   Exercises

10.1   Reliability model of service-based health system: level H

Consider the reliability model of the service-based health system as shown in Figure 10.2. Assume that a trace of 20 runtime data points is observed, each representing the system taking a sample of vital parameters. Suppose that the analysis of the data triggers a service to change the drug six times, a service to change the dose of the drug seven times, and invokes an alarm service four times. Also assume that two invocations of the analysis service fail; you can assume that all other service invocations complete successfully. Apply the Bayesian updating rule of the DCTM model as specified in this chapter, and compute new estimates for the relevant transitions. Update the reliability model accordingly and compare its state with the initial state of the model shown in Figure 10.2. Draw conclusions.

10.2   Impact of training time on learning in DeltaIoT: level D

An important parameter that needs to be set for a classifier is the time that is spent on online training. In the test setup of DeltaIoT, the training time corresponds to a number of cycles of the time-synchronized IoT network. Download the DeltaIoT exemplar equipped with a learning module. Select the simple version of DeltaIoT with 15 nodes and a classifier. The default setting of the training time for this configuration is set to 45 runs. Run an experiment over 300 cycles and collect the measurements for packet loss and latency. Change the training time now to 10 cycles. Repeat the experiment and collect measurements. Change the training time now to 100 cycles and run the experiment again. Plot the results for the three different settings of the training time. Compare the results and assess. The DeltaIoT exemplar with a learning module can be downloaded via [213]. For more information about the application of a classifier to DeltaIoT, see [159].

10.3   Learning pipeline for extended version of DeltaIoT: level W
Consider an extended version of DeltaIoT with 37 nodes and three adaptation goals: packet loss, latency, and energy consumption. Design a two-stage learning pipeline that first uses classification to select adaptation options that comply with the two threshold goals for packet loss and latency, followed by another learner that further refines the selected adaptation options based on minimizing energy consumption. For an implementation of the extended version of DeltaIoT, see [213].

10.6   Bibliographic Notes

A comprehensive introductory book on Machine Learning in general is [75]. For a classic book on Bayesian theory, see [29]. For background on stochastic models, such as DTMC and queuing networks, see [188].

I. Epifani et al. introduced KAMI, the pioneering approach to maintaining models at runtime [68]. This approach uses Bayesian estimation to update parameters of a probabilistic model that represent uncertainties in the environment and the system.

Quin et al. described an approach to reduce large adaptation spaces to enhance the analysis in self-adaptive systems [159]. The authors use a classifier to divide the adaptation space of a system with two threshold goals into four classes. The options of the class that are predicted to comply with the two goals represent the relevant adaptation options, which are then analyzed.

P. Jamshidi et al. introduced a fuzzy-learning approach that learns and improves scaling rules of a Cloud infrastructure [107]. Concretely, a fuzzy Q-learner, which is a particular type of reinforcement learner, selects a scaling action based on the current scaling rules, observes the effects (i.e. the reward), and based on that adjusts the rules to improve the auto-scaling.

To conclude, we list a number of other inspiring applications of machine learning to self-adaptation. M. Sommer et al. dynamically optimized the signalization of traffic light controllers by combining an extended classifier with a neural network [182]. T. Zhao et al. proposed a reinforcement learning-based framework for the generation and evolution of adaptation rules of self-adaptive systems [220]. N. Bencomo and L. Garcia-Paucar studied the maintenance of runtime models relying on Partially Observable Markov Decision Processes and Bayesian inference [21]. P. Jamshidi et al. applied machine learning to find Pareto-optimal configurations without the need for exploring every configuration, making planning tractable [108].

11

Maturity of the Field and Open Challenges

Going through the seven waves of research and engineering on self-adaptive systems, which span a period of more than two decades, has been quite a journey. To conclude, we peek into the future of the field and propose a number of research challenges for the next five to ten years. But, before zooming into these challenges, we first analyze how the field has matured over time.

11.1 Analysis of the Maturity of the Field

According to a study of Redwine and Riddle [165], it typically takes 15 to 20 years for a technology or a field to mature and get widely spread. Figure 11.1 shows the six common phases that can be distinguished in the maturation process. We enhance this process with foundations that refer to underlying principles and theories on which the development of a field relies. Pivotal in the early years of the field of self-adaptation have been dynamic architectures with architectural styles, the notion of quiescence, and languages to specify dynamic architectures. Foundational principles of self-adaptation are computational reflection with causality, and utility as the means to measure preferences of options to adapt a managed system. Later developments of the field were based on applying model checking techniques at runtime, and more recently exploiting principles from control theory and machine learning.

11.1.1 Basic Research

In the first phase, *basic research,* the problem of self-adaptation is recognized and its scope and nature is discussed. Key ideas and concepts are investigated that later prove fundamental.

Pivotal in the first phase was IBM's identification of the manageability problems of computing systems that were becoming increasingly more complex, and the vision of autonomic computing as a means to tackle the problem. A landmark was MAPE-K, which defines the principal functions of an autonomic system [112]. In parallel with the industry-driven efforts, researchers emphasized the need for flexible software to handle change, and proposed the concept of a runtime architectural model that enables a system to reason about itself at an appropriate level of abstraction to make the necessary adaptation

An Introduction to Self-Adaptive Systems: A Contemporary Software Engineering Perspective,
First Edition. Danny Weyns.
© 2021 John Wiley & Sons Ltd. Published 2021 by John Wiley & Sons Ltd.

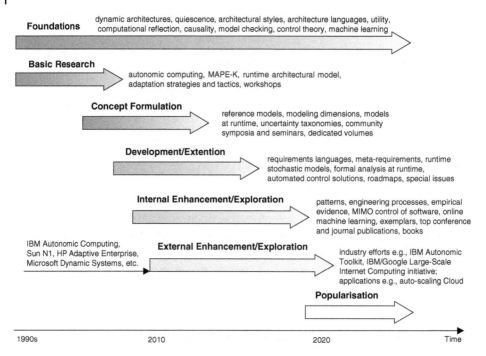

Figure 11.1 Evolution of the maturity of the field of self-adaptation. Gray shading indicates the degree of maturity the field has reached in that phase (phases based on [165]).

decisions [81, 151]. Particularly relevant in the development of the basic ideas and concepts of self-adaptation were the Workshops on Self-Healing Systems[1] and the emergence of the International Conference on Autonomic Computing[2] .

11.1.2 Concept Formulation

In the second phase, *concept formulation,* an informal circulation of ideas coalesces into a compatible set of ideas, and solutions are formulated for specific parts of the problem.

The three-layer model for self-adaptation, which separated change management from goal management, clarified the need for an architectural perspective on the definition of self-adaptive systems [121]. A milestone in the second phase was the Rainbow framework, which demonstrated the practical realization of an external mechanism for self-adaptation [81]. The concept of a model at runtime was instrumental in understanding how to manage the complexity of the design of concrete self-adaptive systems [31]. The classification of modeling dimensions shed light on the distinct facets of what constitutes a self-adaptive software system, establishing a basic vocabulary for specifying the core properties of such systems and selecting suitable solutions [5]. Instrumental in mastering the complex notion of uncertainty was the classification of types and sources of uncertainties [69, 137, 155]. The SEAMS symposium[3] and in particular the series of Dagstuhl

---

1 http://dblp2.uni-trier.de/db/conf/woss/ (11/2019)
2 https://dblp.org/db/conf/icac/index (11/2019)
3 www.hpi.uni-potsdam.de/giese/public/selfadapt/seams/ (11/2019)

seminars[4] on engineering self-adaptive systems have significantly contributed to the maturation of this phase.

11.1.3 Development and Extension

In the third phase, *development and extension,* the principles and concepts of self-adaptation are further clarified and developed, extending the general approach to a broader solution.

Initial applications were developed that demonstrated the usability of self-adaptation. Requirements specification languages for self-adaptive systems [16, 214] and meta-requirements [184] fueled our understanding of the requirements problem of self-adaptation. The need for online reasoning over stochastic models was vital in enhancing our understanding of the decision-making of a self-adaptive system that operates under uncertainty [14, 40, 210]. The Push Button Methodology paved the way for automatically constructed controllers [71, 180], and accelerated the application of control theory to realizing self-adaptation [74, 178]. Landmarks of the third phase were two roadmaps produced by the community on engineering self-adaptive systems [50, 58]. The knowledge and the know-how acquired in the third phase was consolidated in a number of journal special issues, see for instance [31, 207].

11.1.4 Internal Enhancement and Exploration

In phase four, *internal enhancement and exploration,* the solutions to self-adaptation are put in a broader engineering context and applied to real problems in different domains. Primary evidence is derived indicating value, and training is established.

Different initiatives toward creating reusable assets for engineering self-adaptive systems were taken, for instance, the specification of patterns [163, 209] and the creation of reusable infrastructure [46, 167]. The community studied the potential impact on software engineering methodologies in the engineering of self-adaptive systems [6, 42]. First empirical evidence was established to demonstrate the value of self-adaptation. For instance, a controlled experiment was performed to demonstrate that external feedback loops improve the design of self-adaptive systems [208]. Another example empirically validated a goal model to address uncertainty in an industry-provided application handling the dynamic reconfiguration of a remote data mirroring network [78]. To enhance their practicality, initially developed solutions to self-adaptation where enhanced. For instance, control-based solutions were developed to deal with complex types of goals and uncertainties [72, 179], and machine learning was applied to tackle the complexity of large-scale application settings [67, 108, 159]. The establishment of a set of exemplars[5] played a crucial role in the enhancement of the field. Researchers started to present results on self-adaptation at top conferences and published articles in top journals. Courses on self-adaptive systems become part of the main curricula of universities all over the world; a few examples are York University in Canada,[6] which offers a course on Engineering Adaptive Systems, Linnaeus University in Sweden[7] offers a course on Adaptive Software Systems, and the University of Waikato

---

4 www.hpi.uni-potsdam.de/giese/public/selfadapt/dagstuhl-seminars/ (11/2019)
5 www.hpi.uni-potsdam.de/giese/public/selfadapt/exemplars/ (11/2019)
6 https://www.yorku.ca/index.html (11/2019)
7 https://lnu.se/en/ (11/2019)

in New Zealand[8] offers a course on Engineering Self-Adaptive Systems. An increasing number of tutorials are organized at mainstream software engineering venues, for instance at the International Conference on Automated Software Engineering 2015[9] and the International Conference on Software Architecture 2019.[10] The first books appear that provide comprehensive overviews of the field from a particular angle, see for instance [62].

11.1.5 External Enhancement and Exploration

In phase five, *external enhancement and exploration,* a broader community gets engaged in the field, with people from outside the community in which solutions have been developed. This phase is characterized by substantial evidence of value and applicability of the solutions.

The field of self-adaptation is still in an early stage of phase five. Early industrial initiatives centered around autonomic computing include IBM's Autonomic Computing, Sun's N1, HP's Adaptive Enterprise, and Microsoft's Dynamic Systems. In the past decade, various prominent ICT companies have invested significantly in the study and application of self-adaptation [34]. These efforts have resulted in various practical tools, for instance for optimizing energy consumption of a server park and enhancing the capacity efficiency of a Cloud. A characteristic example is Amazon Elastic Compute Cloud Auto Scaling,[11] which offers a Web service that automatically adds or removes resources according to conditions defined by an operator. As such, the service maintains the health and availability of the system by responding to changing demand or based on predicted load of the system. A number of R&D efforts have evaluated the application of self-adaptation beyond mere resource and infrastructure management. For instance, [43] applies self-adaptation to an industrial middleware solution that monitors and manages highly populated networks of devices, [56] applies self-adaptive techniques to role-based access control for business processes, and [211] automates the management of a practical IoT deployment using self-adaptation. These efforts highlight the potential of self-adaptation, but point also to challenges, such as trustworthiness.

11.1.6 Popularization

Finally, the last phase, *popularization,* is characterized by the appearance of production-quality applications. Systems and tools are developed and commercialized that propagate throughout community of users.

The field of self-adaptation is in a very early stage of popularization. Examples of self-adaptation techniques that have found their way to industrial applications are automated server management, Cloud elasticity, and automated data center management as already highlighted above. However, the industrial application of self-adaptation needs to go beyond traditional management of computing infrastructure into other industries and, for instance, facilitate the Industry 4.0 transformation.

---

8 https://www.waikato.ac.nz/ (11/2019)
9 https://ase2015.unl.edu/\LY1\textbackslash#self-adaptive-systems-tutorial (11/2019)
10 https://swk-www.informatik.uni-hamburg.de/\HCode{<SPitie/>}icsa2019/attending/program/index.html\#tutorial2 (11/2019)
11 https://aws.amazon.com/ec2/autoscaling/ (11/2019)

11.1.7 Conclusion

After a relatively slow exploratory start, the maturity of the field of self-adaptation has taken up significantly from around 2006 and is now following the regular path of maturation. The field is currently in the phases of internal and external enhancement and exploration. The wide spread of self-adaptation through practical applications will be of critical importance for the field to reach full maturity.

11.2 Open Challenges

To conclude, we look at open challenges for the next five to ten years. Predicting the future is not easy and not without risk. In the past decade, the community has produced several research roadmaps, see [50], [58], and [59]. These roadmaps resulted from intensive community meetings and provide a wealth of open challenges on a wide variety of aspects of engineering self-adaptive systems. Here we take a different stance and present a set of open research challenges by speculating how the field may evolve in the future based on the seven waves the field went through in the past. We start with a number of short term challenges that fit within current waves. Then we look at challenges in a long term that go beyond the current waves.

11.2.1 Challenges Within the Current Waves

We first look at a set of challenges that go beyond the state of the art within the seven waves of research on self-adaptation.

11.2.1.1 Evidence for the Value of Self-Adaptation

The first wave of research on self-adaptation emerged over two decades ago. At that time, both engineers in industry and academics expressed the necessity for self-adaptation as a means to manage the growing complexity of computing systems. The expectations of self-adaptation were particularly high. Answering the question of whether self-adaptation has achieved its promise, or to what extent this is the case, requires evidence and validation. However, as pointed out by Y. Brun [34] and afterwards confirmed in a systematic literature review from 2013, the validation of contributions to self-adaptation at that time was not done systematically and was often limited to simple example applications [199]. Systematic empirical studies on new research results and industrial validation were rare, and this has not changed so much since. As an illustration, the first systematic empirical evidence for the claimed benefits of external mechanisms for adaptation (in particular, a feedback loop-based managing system) over internal mechanisms was only provided in 2013 [208]. An important challenge for researchers and engineers that crosscuts the different waves will be to develop robust approaches and demonstrate their applicability and value in practice. Essential to that will be to gather empirical evidence based on rigorous methods, in particular controlled experiments and case studies. Initially, such studies can be set up with advanced master students. However, to demonstrate the true value of self-adaptation, it will be essential to involve industry practitioners in such validation efforts.

11.2.1.2 Decentralized Settings

A principal insight of the second wave is that external mechanisms realized through a MAPE-based feedback loop are essential to any self-adaptive system. In a concrete realization of a self-adaptive system, the MAPE functions may directly map to components of the managing system, or they may be integrated. Regardless of the concrete realization, from a conceptual point of view, MAPE takes a centralized perspective on realizing self-adaptation. When systems are large and complex, a single centralized MAPE loop may not be sufficient to realize self-adaptation. A number of researchers have investigated self-adaptation in decentralized settings. For instance, the authors of [209] described a set of patterns in which the functions from multiple MAPE loops coordinate in different ways. R. Calinescu et al. presented a formal approach where MAPE loops coordinate with one another, while providing guarantees for the adaptation decisions they make [41]. M. Caporuscio et al. [45] described a decentralized middleware solution for self-assembling distributed services. While each of these examples take steps toward mastering self-adaptation in decentralized settings, the solutions are tailored to the specific settings and problems they tackle. An open challenge for future research is to study principled solutions to decentralized self-adaptation. This challenge breaks down into several questions that need to be answered: How should the responsibilities of adaptation be divided among the entities? To what extent can the entities trust each other, and how can the necessary trust be established? What knowledge should entities share? What coordination mechanisms can the MAPE functions use to coordinate local adaptations? What interaction protocols do MAPE loops require to realize different types of adaptation goals? How can unwanted emergent behavior be avoided in the decentralized system?

11.2.1.3 Domain-Specific Modeling Languages

Wave three made clear that runtime models play a central role in the realization of self-adaptive systems. A number of languages have been proposed that support the design of models for self-adaptive systems. However, these languages often have a specific focus. An example is Stitch, a language for representing repair strategies within the context of architecture-based adaptation [52]. EUREMA offers a UML-based language that supports the design of feedback loops [195]. These models can be directly deployed and interpreted at runtime to realize adaptation. D. Iglesia et al. presents a set of templates in the form of timed automata that can be used to design and verify the MAPE functions and runtime models of a self-adaptive system [104]. The SCEL language offers a set of primitive abstractions to represent and reason about knowledge and behaviors of autonomic systems [166]. Despite these and similar efforts, the design of most self-adaptive systems currently is based on general purpose modeling paradigms. This may hamper communication among stakeholders, productivity, and reuse. A challenge for future research is to define domain-specific modeling languages that offer first-class support for engineering self-adaptive systems. Such languages would embody domain knowledge, and thus enable the conservation, reuse, and further development of knowledge and know-how. Contrary to traditional systems, where models are primarily design-time artifacts, in self-adaptive systems models are in particular runtime artifacts. We list a few key questions that need to be answered to tackle this challenge: What are the key domain abstractions that need to be supported by modeling languages for self-adaptive systems? How can one support the seamless integration of design time modeling (human-driven) with runtime use of models (machine-driven)? What is the required degree of formality of domain specific modeling languages for self-adaptive systems? How can one measure the effectiveness of a domain specific modeling language for self-adaptive systems?

11.2.1.4 Changing Goals at Runtime

One of the key insights of the fourth wave is the driving role of goals for self-adaptation and their operationalization as first class runtime entities. In the three-layer model for self-adaptation, reasoning about the feasibility of achieving the goals under changing conditions, and if necessary adapting the means for achievement or changing the goals, is a responsibility of the goal management layer. However, goal management has received little attention in research so far. Goal management is basically limited to the representation of goals as first-class entities, which support the decision-making of adaptation under uncertainty. A typical example is presented in [20], where goal realization strategies are associated with decision alternatives, and reasoning about partial satisfaction of goals is based on probabilistic models. Approaches have been developed that support the dynamic selection of goals. For instance, Shevtsov et al. described a control-based approach that allows adaptation goals to be activated, deactivated, and adjusted on the fly based on conditions that are defined before deployment but that can only be resolved during operation [179]. An example of an approach that supports on-the-fly updates of goals and the corresponding MAPE functions is described in [202], but this approach requires manual intervention. The engineer needs to design and verify the updated models offline before they can be deployed, replacing the running models. A challenge for future research is to support changing goals at runtime automatically, in particular support for dynamically adding new goals. Changing goals is particularly challenging and raises several fundamental questions that need to be answered: How to automatically detect the need for changing goals, which include both changes imposed by operators and changes required based on dynamics in the system or its operational context? How to enhance runtime goal models with first class support for dynamic changes of goals? How to support the online synthesis of new adaptation functions according to the changing goals? How to ensure that the new adaptation functions comply with the changing goals? How to enact new adaptation functions and synchronize the updates with the ongoing activities of the self-adaptive system?

11.2.1.5 Complex Types of Uncertainties

Wave five made clear that handling uncertainty is a key driver for the application of self-adaptation. The focus of research in self-adaptation so far has primarily been on parametric uncertainties, i.e. the uncertainties related to the values of model elements that are unknown. A typical example is a Discrete Time Markov Model that represents uncertainties as probabilities of transitions between states [39]. The model parameters are dynamically updated using Bayesian estimation. Similarly, D. Cooray et al. keep a reliability model updated using dynamic learning with Hidden Markov Models [55]. N. Esfahani et al., on the other hand, rely on possibility theory to make the trade-off between different configuration alternatives [70]. A challenge for future research is to support self-adaptation for complex types of uncertainties. One particular type is structural uncertainty, i.e. uncertainty related to the inability to accurately model real-life phenomena. Structural uncertainty may for instance occur in large-scale Internet-of-Things applications, where up-to-date knowledge of the complete network may not be available. Another domain in which structural uncertainty may occur is Cyber Physical Systems. In this domain, structural uncertainty may result from the heterogeneity of these types of systems and their inherent open-endedness. Structural uncertainties may manifest themselves as model inadequacy, model bias, model discrepancy, etc. To tackle this problem, techniques from other fields may provide a starting point. E.g. in health economics, techniques such as model averaging and discrepancy modeling have been used to deal with structural uncertainties [32]. Tackling the problem of managing structural uncertainty raises several questions that need to be answered: How to represent structural uncertainty? How to collect data relevant to structural uncertainty? How to resolve structural uncertainty enabling decision-making about adaptations? When and how to involve humans in mitigating uncertainty in situations that were unforeseeable?

11.2.1.6 Control Properties versus Quality Properties

The sixth wave showed that control theory offers a mathematical basis for providing guarantees for applications that use control-based software adaptation. These guarantees concern static properties, in particular stability, and dynamic properties, including settling time, overshoot, and steady state error [73, 74]. On the other hand, stakeholders of self-adaptive software systems typically express adaptation requirements in the form of non-functional properties, such as performance, reliability, and cost of the system, see for instance [191]. Given the differences between the paradigms of control theory on the one hand and traditional software engineering on the other hand, it is not easy to reconcile these two types of properties. Consequently, it is not clear how the guarantees of control properties translate to guarantees of non-functional properties (and vice versa). A challenge for future research is to define a (bi-directional) mapping between control properties and software quality properties. Tackling this challenge raises several questions that need to be answered: What are the key properties in control theory that are relevant to control-based software adaptation? What are the key quality properties from a software engineering perspective? How can the properties in both worlds be characterized enabling a mapping between the two? What are the commonalities, differences, and potential complementarities among the different properties? How can the mapping between the properties be evaluated?

11.2.1.7 Search-based Techniques

The seventh wave provides initial evidence for the usefulness of machine learning techniques to enhance the different functions that are required to realize self-adaptation. Search-based techniques offer an interesting related and complementary approach to addressing complex problems in self-adaptation. Search-based techniques are attractive because they offer a suite of solutions to tackle large complex problem spaces with multiple competing and conflicting objectives [93]. Examples of such techniques are multi-objective optimization and evolutionary computing. These techniques have demonstrated their value in software engineering in general [91]. Addressing uncertainty in self-adaptive system using evolutionary-based techniques has the benefit that they are not dependent on training data. To deal with multiple interrelated, difficult-to measure, and evolving quality properties, Z. Coker et al. argued for the application of stochastic search techniques to self-adaptive systems – techniques such as hill climbing and genetic algorithms, which incorporate an element of randomness. These techniques are well-suited to handling multi-dimensional search spaces and complex problems, situations that often apply to self-adaptive systems [54]. Two examples of initial work in this field follow. A. Ramirez et al. applied novelty search, an evolutionary-based search technique, to automatically discover environmental conditions that lead to requirements violations of a self-adaptive system. Such conditions facilitate the identification of goals that are not mitigated properly – information that can be used to prevent unwanted behaviors [164]. C. Kinneer et al. presented a planner based on genetic programming that reuses existing plans. The authors proposed a series of techniques to lower the costs of reuse, allowing genetic techniques to leverage existing information to improve planning utility when re-planning for unexpected changes [117]. With the increasing complexity of systems, a challenge for future research is to investigate search-based techniques for various aspects of self-adaptive systems, from testing to model and code evolution as well as generation, verification and assurances, among others.

11.2.2 Challenges Beyond the Current Waves

To conclude, we speculate on a number of challenges in the long term that may trigger new waves of research in the field of self-adaptation.

11.2.2.1 Exploiting Artificial Intelligence

Artificial intelligence (AI) offers systems the ability to make decisions, learn, and improve in order to perform complex tasks [169]. The field of AI covers a broad range of sub-fields, ranging from expert systems and decision-support systems to multi-agent systems, computer vision, natural language processing, speech recognition, machine learning, neural networks and deep learning, and cognitive computation, among others. Areas in which AI techniques have been shown to be useful in software engineering in general are probabilistic reasoning, learning and prediction, and computational search [92]. A number of AI techniques have been turned into mainstream technology, such as machine learning and data analytics. Other techniques are still in a development phase – examples are natural language processing and online reasoning. AI techniques have the potential to disruptively accelerate the capabilities of self-adaptation. AI techniques can play a central role in virtually every stage of adaptation, from defining self-adaptive systems to processing large amounts of data, performing smart analysis and man–machine co-decision making, to coordinating adaptations in large-scale decentralized systems. Realizing the full potential of AI for self-adaptation poses a variety of challenges, including advancing AI techniques, making the techniques secure and trustworthy, and providing solutions for controlling or predicting the behavior of systems that are subject to continuous change. An important remark of P. Norvig in this context is that AI programs are different. One of the key differences is that traditional software essentially hides uncertainty, while AI techniques are fundamentally dealing with uncertainty [147], which emphasizes the potential of AI techniques for self-adaptive systems. Hence, an interesting research area is the exploitation of AI techniques in the realization of self-adaptive systems.

11.2.2.2   Dealing with Unanticipated Change

So far, software is, in essence, a work product of human efforts. Ultimately, a computing machine can only execute what humans have designed and programmed. Nevertheless, recent advances have demonstrated that machines equipped with software can have incredible capabilities to make decisions in complex problems; examples are machines participating in complex strategic games such as chess and cars that drive autonomously. Such examples raise the intriguing question of the extent to which we can develop software that will be able to handle conditions that were not (fully) anticipated at the time when the software was developed. A characteristic example is cyber-attacks, which have become increasingly sophisticated and are often very difficult to anticipate [216]. Applied to the field of self-adaptation, this paradoxical question raises the interesting research problem of how to deal with unanticipated change. A traditional perspective on tackling this problem is to seamlessly integrate adaptation (i.e. the continuous machine-driven process of self-adaptation to deal with known unknowns) with evolution (i.e. the continuous human-driven process of updating the system to deal with unknown unknowns). This idea goes back to the pioneering work of Oreizy et al. [151]. Realizing this idea will require bridging the gap between self-adaptation and software evolution, which aligns with the evolving application of continuous development, where software is produced in short cycles and delivered through automated deployments [175]. A radically different perspective on tackling this problem would be to conceive a system as a dynamic composition of learning processes. The idea would then be to enhance the system with self-learning capabilities. This will enable the system to learn from the data it collects and autonomously evolve its own learning processes under changing and unforeseen conditions. For inspiration, see for instance [221], where the authors propose deep meta-learning and demonstrate its benefits on image recognition problems.

11.2.2.3   Trust and Humans in the Loop

Self-adaptive software is increasingly permeating a variety of domains, including manufacturing, healthcare, and finance. In many of these domains the software is expected to meet strict requirements. Hence stakeholders require the self-adaptive systems to be trustworthy, i.e. they want guarantees that the system operates correctly at all times. In the context of an R & D project on automating the management of an industrial IoT system, stakeholders emphasized that replacing experienced operators by equipping the system with self-adaptation capabilities requires trust and faith in the outcome the automated decision making [211]. While stakeholders may trust the system based on evidence obtained within the anticipated envelope of self-adaptive behavior, concerns are raised about the effects on the system when abnormal conditions occur. An interesting area for further research is establishing trust in self-adaptive systems. We highlight three possible angles from which to start tackling this problem. One approach is to establish trust is using sound technical solutions. An example of a pioneering approach is ENTRUST, which offers an end-to-end methodology for developing trustworthy self-adaptive software [42]. ENTRUST combines design-time and runtime modelling and verification with industry-adopted assurance processes based on dynamic assurance cases to offer trustworthiness. Blockchains are another technology that can be exploited to establish trust. The central abstraction of blockchain systems is a ledger – an indelible, append-only log of transactions that take place between various parties [38, 95]. Placing transactions on a blochchain is based on achieving consensus. Ledgers must be tamper-proof: no party can add, delete, or modify ledger entries once they have been recorded. The potential of blockchains applies in particular to self-adaptive systems in decentralized settings, where the distributed ledger offers a tamper-proof repository of transactional adaptations agreed among the entities. As such the blockchain achieves trustworthiness between interacting entities without the need to explicitly establish such trust. One approach to tackle the trust issue regarding abnormal conditions is using anomaly detection techniques [48]. Such techniques collect and process data during operation in order to detect outliers and anomalies that may be unknown a priori. This provides a basis for predicting patterns and trends and warn stakeholders in case of critical events that fall outside the regular envelope of system behavior. In general, anomaly detection largely depends on the system and context at hand. A more fundamental solution to establishing trust could be involving the human in the self-adaptation loop. To provide trusted services in the presence of changes, J. Camara et al. argued that self-adaptive systems can benefit from incorporating human input into the decision-making process to provide better insight into the best way of adapting the system, or employing humans as system-level effectors to execute adaptations, for instance as a fallback mechanism [44]. Incorporating humans in self-adaptation is challenging and demands new approaches to systematically reasoning about the behavior of human participants as well as man–machine interaction mechanisms. While any of the proposed techniques may contribute to establishing trust in self-adaptive systems, practical solutions may require a combination of technical solutions with active involvement of humans in the adaptation workflow.

11.2.2.4 Ethics for Self-Adaptive Systems

Software systems are playing an increasingly dominating role in industry, government, health, transportation, education, and many other domains of our society. Because software is developed by software engineers, they have significant opportunities to do good or cause harm, directly or indirectly. To ensure that software will be used for good and make the world a better place, software engineers should adhere to ethical principles. A joint task force of IEEE and ACM has brought these principles together in a "Code of Ethics" [149]. The items numbered from 1 to 8 in Figure 11.2 give an overview of these principles.

Principle 1 states that software engineers shall act consistently with the public interest. Principle 2 states that software engineers shall act in a manner that is in the best interests of their clients, consistent with the public interest. The third principle states that software engineers shall ensure that their products and related modifications meet the highest professional standards possible. Principle 4 states that software engineers shall maintain integrity and independence in their professional judgment, while principle 5 states that software engineering professionals shall subscribe to and promote an ethical approach to the management of software development and maintenance. Principle 6 states that software engineers shall advance the integrity and reputation of the profession consistent with the public interest. Principle 7 states that software engineers shall be fair and supportive to their colleagues. Finally, principle 8 states that software engineers shall participate in lifelong learning regarding the practice of their profession.

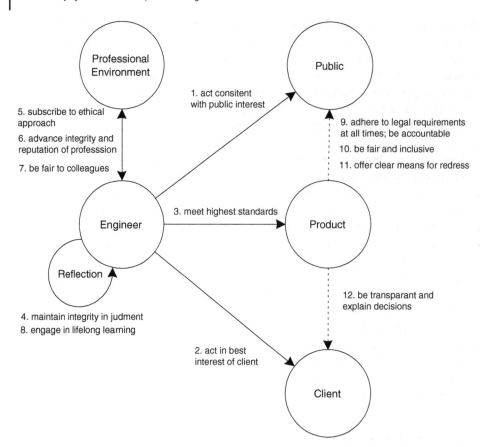

Figure 11.2 Schematic overview of the Code of Ethics of IEEE/ACM; circles represent actors, arrows represent ethical principles. Principles 1 to 8 apply to the basic Code of Ethics. Principles 9 to 12 suggest new ethical principles for autonomous systems, such as self-adaptive systems.

With the emergence of computing systems that take decisions on behalf of humans, new additional ethical principles will be required, as endorsed by several initiatives all over the world. For example, the United States' Fair Credit Reporting Act and European Union's General Data Protection Regulation (GDPR) prescribe that data must be processed in a way that is fair and unbiased. GDPR also alludes to the right of an individual to receive an explanation about decisions made by an automated system. Another example is the recommendation provided by The World Economic Forum to prevent discriminatory outcomes by machine learning applications [76]. Researchers are also taking initiatives toward the definition of ethical frameworks, for instance for the design of autonomous intelligent systems [125]. Since self-adaptive systems are characterized by autonomy and these systems take decisions on behalf of humans, the need for new principles applies to these systems.

The items numbered from 9 to 12 in Figure 11.2 suggest an initial set of new ethical principles for autonomous systems. Whereas the basic principles (1 to 8) directly apply to the engineers of software systems, the new principles are exposed through the systems themselves, and apply indirectly to the engineers that developed the systems. Hence it is the responsibility of the engineer to ensure that the system complies with all new principles. For principles 9 and 10, agreement is growing among various parties. Principles 11 and 12, on the other hand, are more speculative and subject of debate.

Principle 9 states that the system shall adhere to legal requirements at all times and be accountable for that. Principle 10 states that the system shall be fair and explicitly take fairness into account in its evaluation metrics. Principle 11 states that the system shall offer clear means of redress for those affected by disparate impacts of it. Finally, principle 12 states that the system shall be transparent and explain the decisions it makes in an understandable way to those that are affected by the decision-making.

This initial set of ethical principles offer inspiration for extending the Code of Ethics for autonomous and self-adaptive systems. Establishing such an extended code is complex and will require a huge concerted effort from a broad range of actors. Not only is there a need for an agreement on new ethical principles, but these principles need to be implemented, which raises numerous technical challenges. For instance, approving that autonomous and self-adaptive systems are safe and meet their specifications is particularly challenging since such systems operate under uncertainty that is often difficult to anticipate. When a code ultimately expresses the consensus of a professional community on ethical issues, it will offer a means to educate the public that will be affected by autonomous software systems, but also to teach new generations of software engineers about the ethical obligations in their profession.

11.3 Epilogue

In a world where computing systems are rapidly converging into large open ecosystems, automation is inevitable and so is the uncertainty systems are exposed to. This applies to many systems we build today and it will be a dominating factor of most systems we will build in the future. The challenges software engineers face to build high-quality software that is able to handle uncertainty are huge. Self-adaptation has enormous potential to tackle many of these challenges. The field has come a long way, as testified in this book, and a substantial body of knowledge has been developed over the past two decades. Building upon these established foundations, addressing key challenges now requires consolidating the knowledge and turning results into robust and reusable solutions. This will move the field forward and propagate the technology throughout a broad community of users in practice. Tackling these challenges is not without risk as it requires researchers to leave their comfort zone and expose their research results to the complexity of practical systems. However, taking this risk will propel diffusion of the technology, open new opportunities for research, and pave the way for self-adaptation to reach full maturity as a discipline.

Bibliography

1 T. Abdelzaher, Y. Diao, J. Hellerstein, C. Lu, and X. Zhu. Introduction to Control Theory And Its Application to Computing Systems. In Zn Liu and C. Xia, editors, *Performance Modeling and Engineering*, pages 185–215. Springer, 2008. ISBN 978-0-387-79361-0. doi: 10.1007/978-0-387-79361-0_7. URL https://doi.org/10.1007/978-0-387-79361-0_7.

2 OSGI Alliance. *OSGI Service Platform, Release 3*. IOS Press, Inc., 2003. ISBN 978-1-58603-311-8. URL https://www.iospress.nl/book/osgi-service-platform-release-3/.

3 R. Alur and D. Dill. A Theory of Timed Automata. *Theoretical Computer Science*, 126(2):183–235, 1994. ISSN 0304-3975. doi: https://doi.org/10.1016/0304-3975(94)90010-8. URL http://www.sciencedirect.com/science/article/pii/0304397594900108.

4 S. Aminikhanghahi and D. Cook. A Survey of Methods for Time Series Change Point Detection. *Knowledge Information Systems*, 51(2):339–367, May 2017. ISSN 0219-1377. doi: 10.1007/s10115-016-0987-z. URL https://doi.org/10.1007/s10115-016-0987-z.

5 J. Andersson, R. de Lemos, S. Malek, and D. Weyns. Modeling Dimensions of Self-Adaptive Software Systems. In B. Cheng, R. de Lemos, H. Giese, P. Inverardi, and J. Magee, editors, *Software Engineering for Self-Adaptive Systems*, pages 27–47. Springer, 2009. ISBN 978-3-642-02161-9. doi: 10.1007/978-3-642-02161-9_2. URL http://dx.doi.org/10.1007/978-3-642-02161-9_2.

6 J. Andersson, L. Baresi, N. Bencomo, R. de Lemos, A. Gorla, P. Inverardi, and T. Vogel. Software Engineering Processes for Self-Adaptive Systems. In R. de Lemos, H. Giese, H. Müller, and M. Shaw, editors, *Software Engineering for Self-Adaptive Systems II*, pages 51–75. Springer, 2013. ISBN 978-3-642-35813-5. doi: 10.1007/978-3-642-35813-5_3. URL https://doi.org/10.1007/978-3-642-35813-5_3.

7 K. Angelopoulos, A. Papadopoulos, V. Silva Souza, and J. Mylopoulos. Engineering Self-Adaptive Software Systems: From Requirements to Model Predictive Control. *Transactions on Autonomous and Adaptive Systems*, 13(1):1:1–1:27, April 2018. ISSN 1556-4665. doi: 10.1145/3105748. URL http://doi.acm.org/10.1145/3105748.

8 K. Astrom and R. Murray. *Feedback Systems: An Introduction for Scientists and Engineers*. Princeton University Press, 11/2019. ISBN 978-0-691-13576-2.

An Introduction to Self-Adaptive Systems: A Contemporary Software Engineering Perspective,
First Edition. Danny Weyns.
© 2021 John Wiley & Sons Ltd. Published 2021 by John Wiley & Sons Ltd.

URL http://www.cds.caltech.edu/~murray/amwiki/index.php/
Second_Edition.

9 M. Autili, P. Inverardi, R. Spalazzese, M. Tivoli, and F. Mignosi. Automated Synthesis of Application-layer Connectors from Automata-based Specifications. *Journal of Computer and System Sciences*, 104:17–40, September 2019. doi: 10.1016/j.jcss.2019.03.001. URL https://www.sciencedirect.com/science/article/pii/
S0022000019300248.

10 C. Baier and J.P. Katoen. *Principles of Model Checking*. The MIT Press. Cambridge, Massachusetts, 2008. ISBN 9780262026499. URL https://mitpress.mit.edu/
books/principles-model-checking.

11 D. Barbosa, R. de Moura Lima, P. Maia, and E. Junior. Lotus@Runtime: A Tool for Runtime Monitoring and Verification of Self-adaptive Systems. In *12th International Symposium on Software Engineering for Adaptive and Self-Managing Systems*, pages 24–30, Buenos Aires, Argentina, 2017. IEEE. ISBN 978-1-5386-1550-8. doi: 10.1109/SEAMS.2017.18. URL https://doi.org/10.1109/SEAMS.2017.18.

12 D. Barbosa, R. de Moura Lima, P. Maia, and E. Junior. Lotus@Runtime Artifact, November 2019. URL https://www.hpi.uni-potsdam.de/giese/public/
selfadapt/exemplars/lotusruntime/.

13 D. Barbosa, R. de Moura Lima, P. Maia, and E. Junior. *Lotus@Runtime Github*, November 2019. URL https://github.com/davimonteiro.

14 L. Baresi and C. Ghezzi. The Disappearing Boundary between Development-time and Run-time. In *Future of Software Engineering Research*, pages 17–22, Santa Fe, New Mexico, USA, 2010. doi: 10.1145/1882362.1882367. URL https://doi.org/10
.1145/1882362.1882367.

15 L. Baresi, D. Bianculli, C. Ghezzi, S. Guinea, and P. Spoletini. Validation of Web Service Compositions. *IET Software*, 1(6):219–232, December 2007. ISSN 1751-8814. doi: 10.1049/iet-sen:20070027. URL https://ieeexplore.ieee.org/document/
4435102.

16 L. Baresi, L. Pasquale, and P. Spoletini. Fuzzy Goals for Requirements-Driven Adaptation. In *18th IEEE International Requirements Engineering Conference*, pages 125–134, Banff, Alberta, Canada, 2010. IEEE. ISBN 978-0-7695-4162-4. doi: 10.1109/RE.2010.25. URL http://dx.doi.org/10.1109/RE.2010.25.

17 C. Barna, H. Ghanbari, M. Litoiu, and M. Shtern. Hogna: A Platform for Self-adaptive Applications in Cloud Environments. In *10th International Symposium on Software Engineering for Adaptive and Self-Managing Systems*, pages 83–87, Florence, Italy, 2015. IEEE. doi: 10.1109/SEAMS.2015.26. URL http://dl.acm.org/citation.cfm?
id=2821357.2821372.

18 C. Barna, H. Ghanbari, M. Litoiu, and M. Shtern. Hogna Artifact, November 2019. URL https://www.hpi.uni-potsdam.de/giese/public/selfadapt/
exemplars/hogna/.

19 J. Beal, M. Viroli, D. Pianini, and F. Damiani. Self-Adaptation to Device Distribution in the Internet of Things. *ACM Transactions on Autonomous and Adaptive Systems*, 12(3):12:1–12:29, September 2017. ISSN 1556-4665. doi: 10.1145/3105758. URL http://doi.acm.org/10.1145/3105758.

20 N. Bencomo and A. Belaggoun. Supporting Decision-Making for Self-Adaptive Systems: From Goal Models to Dynamic Decision Networks. In *International Working Conference on Requirements Engineering: Foundation for Software Quality*, Essen, Germany, 2013. Springer. ISBN 978-3-642-37422-7. doi: 10.1007/978-3-642-37422-7_16. URL `https://link.springer.com/chapter/10.1007/978-3-642-37422-7_16`.

21 N. Bencomo and L. Hernán García Paucar. RaM: Causally-Connected and Requirements-Aware Runtime Models using Bayesian Learning. In *22nd International Conference on Model Driven Engineering Languages and Systems*, pages 216–226, Munich, Germany, 2019. IEEE. doi: 10.1109/MODELS.2019.00005. URL `https://doi.org/10.1109/MODELS.2019.00005`.

22 N. Bencomo, J. Whittle, P. Sawyer, A. Finkelstein, and E. Letier. Requirements Reflection: Requirements as Runtime Entities. In *32nd International Conference on Software Engineering*, Cape Town, South Africa, 2010. IEEE. doi: 10.1145/1810295.1810329. URL `https://ieeexplore.ieee.org/document/6062159`.

23 N. Bencomo, R. France, B. Cheng, and U. Assmann. *Models@Run.Time: Foundations, Applications, and Roadmaps*, volume 8378 of *Lecture Notes in Computer Science*. Springer, 2014. ISBN 978-3-319-08915-7. doi: 10.1007/978-3-319-08915-7. URL `https://www.springer.com/gp/book/9783319089140`.

24 N. Bencomo, S. Götz, and H. Song. Models@Run.Time: A Guided Tour of the State of the Art and Research Challenges. *Software & Systems Modeling*, 18(5):3049–3082, October 2019. doi: 10.1007/s10270-018-00712-x. URL `https://doi.org/10.1007/s10270-018-00712-x`.

25 A. Bennaceur, C. McCormick, J. Galán, C. Perera, A. Smith, A. Zisman, and B. Nuseibeh. Feed Me, Feed Me: An Exemplar for Engineering Adaptive Software. In *11th International Symposium on Software Engineering for Adaptive and Self-Managing Systems*, pages 89–95, Austin, Texas, 2016. ACM. ISBN 978-1-4503-4187-5. doi: 10.1145/2897053.2897071. URL `http://doi.acm.org/10.1145/2897053.2897071`.

26 A. Bennaceur, C. McCormick, J. Galán, C. Perera, A. Smith, Andrea Z., and B. Nuseibeh. Feed Me, Feed Me Artifact, November 2019. URL `https://www.hpi.uni-potsdam.de/giese/public/selfadapt/exemplars/feed-me-feed-me/`.

27 A. Bennaceur, C. McCormick, J. Galán, C. Perera, A. Smith, Andrea Z., and B. Nuseibeh. Feed Me, Feed Me Artifact Website, November 2019. URL `http://sead1.open.ac.uk/fmfm/`.

28 K. Bennett and V. Rajlich. Software Maintenance and Evolution: A Roadmap. In *The Future of Software Engineering*, pages 73–87, Limerick, Ireland, 2000. ACM. ISBN 1-58113-253-0. doi: 10.1145/336512.336534. URL `http://doi.acm.org/10.1145/336512.336534`.

29 J. Bernardo and A. Smith. *Bayesian Theory, 2nd Edition*. Wiley, 2007. ISBN 978-0470028735. URL `https://www.wiley.com/en-us/Bayesian+Theory-p-9780471924166`.

30 D. Berry, B. Cheng, and J. Zhang. The Four Levels of Requirements Engineering for and in Dynamic Adaptive Systems. In *International Workshop on the Design and Evolution of Autonomic Application Aoftware*, Saint Louis, MO, USA, 2005. doi:

10.1109/ICSE.2005.1553669. URL https://ieeexplore.ieee.org/document/ 1553669.

31 G. Blair, N. Bencomo, and R. France. Models@Run.Time. *IEEE Computer*, 42(10):22–27, October 2009. ISSN 0018-9162. doi: 10.1109/MC.2009.326. URL https://dl.acm.org/doi/10.1109/MC.2009.326.

32 L. Bojke, K. Claxton, M. Sculpher, and Palmer S. Characterizing Structural Uncertainty in Decision Analytic Models: A Review and Application of Methods. *Value Health*, 12(5):739–749, July-August 2009. doi: 10.1111/j.1524-4733.2008.00502.x. URL https://onlinelibrary.wiley.com/doi/full/10.1111/j.1524-4733 .2008.00502.x.

33 V. Braberman, N. D'Ippolito, N. Piterman, D. Sykes, and S. Uchitel. Controller Synthesis: From Modelling to Enactment. In *35th International Conference on Software Engineering*, pages 1347–1350, San Francisco, CA, USA, 2013. IEEE. ISBN 978-1-4673-3076-3. doi: 10.1109/ICSE.2013.6606714. URL http://dl.acm.org/ citation.cfm?id=2486788.2487002.

34 Y. Brun. Improving Impact of Self-adaptation and Self-management Research Through Evaluation Methodology. In *International Symposium on Software Engineering for Adaptive and Self-Managing Systems*, Cape Town, South Africa, 2010. ACM. doi: 10.1145/1808984.1808985. URL http://doi.acm.org/10.1145/1808984 .1808985.

35 Y. Brun, G. Marzo Serugendo, C. Gacek, H. Giese, H. Kienle, M. Litoiu, H. Müller, M. Pezzè, and M. Shaw. Engineering Self-Adaptive Systems Through Feedback Loops. In B. Cheng, R. Lemos, H. Giese, P. Inverardi, and J. Magee, editors, *Software Engineering for Self-Adaptive Systems*, pages 48–70. Springer, 2009. ISBN 978-3-642-02160-2. doi: 10.1007/978-3-642-02161-9_3. URL http://dx.doi.org/10.1007/978-3-642- 02161-9_3.

36 Y. Brun, R. Desmarais, K. Geihs, M. Litoiu, A. Lopes, M. Shaw, and M. Smit. A Design Space for Self-Adaptive Systems. In R. de Lemos, H. Giese, H. Muller, and M. Shaw, editors, *Software Engineering for Self-Adaptive Systems II*, pages 33–50. Springer, 2013. ISBN 978-3-642-35813-5. doi: 10.1007/978-3-642-35813-5_2. URL http://dx.doi .org/10.1007/978-3-642-35813-5_2.

37 T. Bures, F. Plasil, M. Kit, P. Tuma, and N. Hoch. Software Abstractions for Component Interaction in the Internet of Things. *Computer*, 49(12): 50–59, December 2016. ISSN 1558-0814. doi: 10.1109/MC.2016.377. URL https://ieeexplore.ieee .org/document/7756271.

38 C. Cachin. Architecture of the Hyperledger Blockchain Fabric. In *Distributed Cryptocurrencies and Consensus Ledgers*, Chicago, Illinois, USA, 2016. IBM Research. URL https://www.zurich.ibm.com/dccl/papers/cachin_dccl.pdf.

39 R. Calinescu, L. Grunske, M. Kwiatkowska, R. Mirandola, and G. Tamburrelli. Dynamic QoS Management and Optimization in Service-Based Systems. *IEEE Transactions on Software Engineering*, 37 (3):387–409, May 2011. ISSN 0098-5589. doi: 10.1109/TSE.2010.92. URL http://dx.doi.org/10.1109/TSE.2010.92.

40 R. Calinescu, C. Ghezzi, M. Kwiatkowska, and R. Mirandola. Self-adaptive Software Needs Quantitative Verification at Runtime. *Communications of the ACM*, 55(9):69–77,

2012. doi: 10.1145/2330667.2330686. URL https://doi.org/10.1145/2330667
.2330686.

41 R. Calinescu, S. Gerasimou, and A. Banks. Self-adaptive Software with Decen-
tralised Control Loops. In *International Conference on Fundamental Approaches
to Software Engineering*, London, UK, 2015. Springer. ISBN 978-3-662-46675-9. doi:
10.1007/978-3-662-46675-9_16. URL https://doi.org/10.1007/978-3-662-
46675-9_16.

42 R. Calinescu, D. Weyns, S. Gerasimou, M. U. Iftikhar, I. Habli, and T. Kelly. Engineering Trustworthy Self-Adaptive Software with Dynamic Assurance Cases. *IEEE
Transactions on Software Engineering*, 44(11):1039–1069, November 2018. ISSN
2326-3881. doi: 10.1109/TSE.2017.2738640. URL https://ieeexplore.ieee.org/
document/8008800.

43 J. Cámara, P. Correia, R. de Lemos, D. Garlan, P. Gomes, B. Schmerl, and R.
Ventura. Evolving an Adaptive Industrial Software System to Use Architecture-based
Self-adaptation. In *8th International Symposium on Software Engineering for Adaptive and Self-Managing Systems*, pages 13–22, San Francisco, CA, USA, 2013. doi:
10.1109/SEAMS.2013.6595488. URL https://doi.org/10.1109/SEAMS.2013
.6595488.

44 J. Camara, G. Moreno, and D. Garlan. Reasoning about Human Participation in
Self-Adaptive Systems. In *10th International Symposium on Software Engineering for
Adaptive and Self-Managing Systems*, pages 146–156, Florence, Italy, 2015. ACM. doi:
10.1109/SEAMS.2015.14. URL https://ieeexplore.ieee.org/abstract/
document/7194669.

45 M. Caporuscio, V. Grassi, M. Marzolla, and R. Mirandola. GoPrime: A Fully Decentralized Middleware for Utility-Aware Service Assembly. *IEEE Transactions on Software
Engineering*, 42(2):136–152, February 2016. doi: 10.1109/TSE.2015.2476797. URL
https://ieeexplore.ieee.org/document/7243346.

46 V. Cardellini, E. Casalicchio, V. Grassi, and F. Lo Presti. Adaptive Management
of Composite Services under Percentile-Based Service Level Agreements. In P.
Maglio, M. Weske, J. Yang, and M. Fantinato, editors, *International Conference on
Service-Oriented Computing*, pages 381–395, San Francisco, CA, USA, 2010. Springer.
ISBN 978-3-642-17358-5. doi: 10.1007/978-3-642-17358-5_26. URL https://link
.springer.com/chapter/10.1007/978-3-642-17358-5_26.

47 C. Cetina, P. Giner, J. Fons, and V. Pelechano. Autonomic Computing through Reuse
of Variability Models at Runtime: The Case of Smart Homes. *Computer*, 42(10):37–43,
October 2009. ISSN 1558-0814. doi: 10.1109/MC.2009.309. URL https://dl.acm
.org/doi/10.1109/MC.2009.309.

48 V. Chandola, A. Banerjee, and V. Kumar. Anomaly Detection: A Survey. *ACM Computing Surveys*, 41(3):15:1–15:58, July 2009. ISSN 0360-0300. doi: 10.1145/1541880.1541882.
URL http://doi.acm.org/10.1145/1541880.1541882.

49 B. Cheng and J. Atlee. Research Directions in Requirements Engineering. In *Future
of Software Engineering*, pages 285–303, Minneapolis, MN, USA, 2007. IEEE. ISBN
0-7695-2829-5. doi: 10.1109/FOSE.2007.17. URL https://doi.org/10.1109/FOSE
.2007.17.

50 B. Cheng, R. de Lemos, H. Giese, P. Inverardi, J. Magee, J. Andersson, B. Becker, N. Bencomo, Y. Brun, B. Cukic, G. Di Marzo Serugendo, S. Dustdar, A. Finkelstein, C. Gacek, K. Geihs, V. Grassi, G. Karsai, H. Kienle, J. Kramer, M. Litoiu, S. Malek, R. Mirandola, H. Müller, S. Park, M. Shaw, M. Tichy, M. Tivoli, D. Weyns, and J. Whittle. Software Engineering for Self-Adaptive Systems: A Research Roadmap. In B. Cheng, R. de Lemos, H. Giese, P. Inverardi, and J. Magee, editors, *Software Engineering for Self-Adaptive Systems*, pages 1–26. Springer, 2009. ISBN 978-3-642-02161-9. doi: 10.1007/978-3-642-02161-9_1. URL https://doi.org/10.1007/978-3-642-02161-9_1.

51 B. Cheng, P. Sawyer, N. Bencomo, and J. Whittle. A Goal-Based Modeling Approach to Develop Requirements of an Adaptive System with Environmental Uncertainty. In *12th International Conference on Model Driven Engineering Languages and Systems*, pages 468–483, Denver, CO, 2009. Springer. ISBN 978-3-642-04424-3. doi: 10.1007/978-3-642-04425-0_36. URL http://dx.doi.org/10.1007/978-3-642-04425-0_36.

52 S. Cheng and D. Garlan. Stitch: A Language for Architecture-based Self-adaptation. *Journal of Systems and Software*, 85(12):2860–2875, December 2012. ISSN 0164-1212. doi: 10.1016/j.jss.2012.02.060. URL http://dx.doi.org/10.1016/j.jss.2012.02.060.

53 S. Cheng and B. Schmerl. ZNN Artifact, November 2019. URL https://www.hpi.uni-potsdam.de/giese/public/selfadapt/exemplars/model-problem-znn-com/.

54 Z. Coker, D. Garlan, and C. Le Goues. SASS: Self-adaptation Using Stochastic Search. In *10th International Symposium on Software Engineering for Adaptive and Self-Managing Systems*, pages 168–174, Florence, Italy, 2015. IEEE. doi: 10.5555/2821357.2821386. URL http://dl.acm.org/citation.cfm?id=2821357.2821386.

55 D. Cooray, S. Malek, R. Roshandel, and D. Kilgore. RESISTing Reliability Degradation Through Proactive Reconfiguration. In *International Conference on Automated Software Engineering*, pages 83–92, Antwerp, Belgium, 2010. ACM. ISBN 978-1-4503-0116-9. doi: 10.1145/1858996.1859011. URL http://doi.acm.org/10.1145/1858996.1859011.

56 C. E. da Silva, J. D. S. da Silva, C. Paterson, and R. Calinescu. Self-Adaptive Role-Based Access Control for Business Processes. In *12th International Symposium on Software Engineering for Adaptive and Self-Managing Systems*, pages 193–203, Buenos Aires, Argentina, 2017. doi: 10.1109/SEAMS.2017.13. URL https://doi.org/10.1109/SEAMS.2017.13.

57 A. David, K. Larsen, and A. Legay et al. Uppaal SMC Tutorial. *International Journal on Software Tools for Technology Transfer*, 17(4): 397–415, 2015. ISSN 1433-2787. doi: 10.1007/s10009-014-0361-y. URL https://doi.org/10.1007/s10009-014-0361-y.

58 R. de Lemos, H. Giese, H. Müller, M. Shaw, J. Andersson, M. Litoiu, B. Schmerl, G. Tamura, N. Villegas, T. Vogel, D. Weyns, L. Baresi, B. Becker, N. Bencomo, Y. Brun, B. Cukic, R. Desmarais, S. Dustdar, G. Engels, K. Geihs, K. Göschka, A. Gorla, V. Grassi, P. Inverardi, G. Karsai, J. Kramer, A. Lopes, J. Magee, S. Malek, S. Mankovski,

R. Mirandola, J. Mylopoulos, O. Nierstrasz, M. Pezzè, C. Prehofer, W. Schäfer, R. Schlichting, D. Smith, J. Pedro Sousa, L. Tahvildari, K. Wong, and J. Wuttke. Software Engineering for Self-Adaptive Systems: A Second Research Roadmap. In R. de Lemos, H. Giese, H. Muller, and M. Shaw, editors, *Software Engineering for Self-Adaptive Systems II*, pages 1–32. Springer, 2010. doi: 10.1007/978-3-642-35813-5_1. URL https://doi.org/10.1007/978-3-642-35813-5_1.

59 R. de Lemos, D. Garlan, C. Ghezzi, H. Giese, J. Andersson, M. Litoiu, B. Schmerl, D. Weyns, L. Baresi, N. Bencomo, Y. Brun, J. Camara, R. Calinescu, M. Cohen, A. Gorla, V. Grassi, L. Grunske, P. Inverardi, J. Jezequel, S. Malek, R. Mirandola, M. Mori, H. Müller, R. Rouvoy, C. Rubira, E. Rutten, M. Shaw, G. Tamburrelli, G. Tamura, N. Villegas, T. Vogel, and F. Zambonelli. Software Engineering for Self-Adaptive Systems: Research Challenges in the Provision of Assurances. In R. de Lemos, D. Garlan, C. Ghezzi, and H. Giese, editors, *Software Engineering for Self-Adaptive Systems III. Assurances*, pages 3–30. Springer, 2017. ISBN 978-3-319-74183-3. doi: 10.1007/978-3-319-74183-3_1. URL https://link.springer.com/chapter/10.1007/978-3-319-74183-3_1.

60 T. De Wolf and T. Holvoet. Emergence Versus Self-Organisation: Different Concepts but Promising When Combined. In *Engineering Self-Organising Systems: Methodologies and Applications*, pages 1–15, Utrecht, The Netherlands, 2005. Springer. ISBN 978-3-540-31901-6. doi: 10.1007/11494676_1. URL http://dx.doi.org/10.1007/11494676_1.

61 A. Dey. Context-Aware Computing. In J. Krumm, editor, *Ubiquitous Computing Fundamentals*, pages 321–352. Chapman & Hall/CRC, 2009. ISBN 1420093606. URL https://dl.acm.org/doi/book/10.5555/1803789.

62 A. Diaconescu, J. McCann, and P. Lalanda. *Autonomic Computing: Principles, Design and Implementation*. Springer, 2013. ISBN 978-1-4471-5007-7. doi: 10.1007/978-1-4471-5007-7. URL https://www.springer.com/gp/book/9781447150060.

63 N. D'Ippolito, V. Braberman, J. Kramer, J. Magee, D. Sykes, and S. Uchitel. Hope for the Best, Prepare for the Worst: Multi-tier Control for Adaptive Systems. In *36th International Conference on Software Engineering*, pages 688–699. ACM, 2014. ISBN 978-1-4503-2756-5. doi: 10.1145/2568225.2568264. URL http://doi.acm.org/10.1145/2568225.2568264.

64 DistriNet. PacketWorld Test Bed, November 2019. URL https://sourceforge.net/projects/packet-world/.

65 S. Dobson, S. Denazis, D. Fernández, A. and Gaïti, E. Gelenbe, F. Massacci, P. Nixon, F. Saffre, Ni. Schmidt, and F. Zambonelli. A Survey of Autonomic Communications. *ACM Transactions on Autonomous and Adaptive Systems*, 1(2):223–259, December 2006. ISSN 1556-4665. doi: 10.1145/1186778.1186782. URL http://doi.acm.org/10.1145/1186778.1186782.

66 R. Edwards and N. Bencomo. DeSiRE: Further Understanding Nuances of Degrees of Satisfaction of Non-functional Requirements Trade-off. In *13th International Symposium on Software Engineering for Adaptive and Self-Managing Systems*, pages 12–18, Gothenburg, Sweden, 2018. ACM. ISBN 978-1-4503-5715-9. doi:

10.1145/3194133.3194142. URL http://doi.acm.org/10.1145/3194133
.3194142.

67 A. Elkhodary, N. Esfahani, and S. Malek. FUSION: A Framework for Engineering
Self-tuning Self-adaptive Software Systems. In *18th ACM SIGSOFT International
Symposium on Foundations of Software Engineering*, pages 7–16, Santa Fe, New Mexico, USA, 2010. ACM. ISBN 978-1-60558-791-2. doi: 10.1145/1882291.1882296. URL
http://doi.acm.org/10.1145/1882291.1882296.

68 I. Epifani, C. Ghezzi, R. Mirandola, and G. Tamburrelli. Model Evolution by Run-time
Parameter Adaptation. In *31st International Conference on Software Engineering*, pages
111–121, Vancouver, British Columbia, Canada, 2009. IEEE. ISBN 978-1-4244-3453-4.
doi: 10.1109/ICSE.2009.5070513. URL http://dx.doi.org/10.1109/ICSE.2009
.5070513.

69 N. Esfahani and S. Malek. Uncertainty in Self-Adaptive Software Systems. In R.
de Lemos, H. Giese, H. Müller, and M. Shaw, editors, *Software Engineering for
Self-Adaptive Systems II*, pages 214–238. Springer, 2013. ISBN 978-3-642-35813-5. doi:
10.1007/978-3-642-35813-5_9. URL https://doi.org/10.1007/978-3-642-
35813-5_9.

70 N. Esfahani, E. Kouroshfar, and S. Malek. Taming Uncertainty in Self-adaptive
Software. In *19th SIGSOFT Symposium and the 13th European Conference on Foundations of Software Engineering*, pages 234–244, Szeged, Hungary, 2011. ACM. ISBN
978-1-4503-0443-6. doi: 10.1145/2025113.2025147. URL http://doi.acm.org/10
.1145/2025113.2025147.

71 A. Filieri, H. Hoffmann, and M. Maggio. Automated Design of Self-adaptive Software with Control-theoretical Formal Guarantees. In *36th International Conference
on Software Engineering*, pages 299–310, Hyderabad, India, 2014. ACM. ISBN
978-1-4503-2756-5. doi: 10.1145/2568225.2568272. URL http://doi.acm.org/
10.1145/2568225.2568272.

72 A. Filieri, H. Hoffmann, and M. Maggio. Automated Multi-objective Control for
Self-adaptive Software Design. In *10th Joint Meeting on Foundations of Software
Engineering*, pages 13–24, Bergamo, Italy, 2015. ACM. ISBN 978-1-4503-3675-8.
doi: 10.1145/2786805.2786833. URL http://doi.acm.org/10.1145/2786805
.2786833.

73 A. Filieri, M. Maggio, K. Angelopoulos, N. D'Ippolito, I. Gerostathopoulos, A. Hempel,
H. Hoffmann, P. Jamshidi, E. Kalyvianaki, C. Klein, F. Krikava, S. Misailovic, A.
Papadopoulos, S. Ray, A. Sharifloo, S. Shevtsov, M. Ujma, and T. Vogel. Software
Engineering Meets Control Theory. In *10th International Symposium on Software Engi-
neering for Adaptive and Self-Managing Systems*, pages 71–82, Florence, Italy, 2015.
IEEE. doi: 10.1109/SEAMS.2015.12. URL http://dl.acm.org/citation.cfm?
id=2821357.2821370.

74 A. Filieri, M. Maggio, K. Angelopoulos, N. D'ippolito, I. Gerostathopoulos, A. Hempel,
H. Hoffmann, P. Jamshidi, E. Kalyvianaki, C. Klein, F. Krikava, S. Misailovic, A.
Papadopoulos, S. Ray, A. Sharifloo, S. Shevtsov, M. Ujma, and T. Vogel. Control
Strategies for Self-Adaptive Software Systems. *ACM Transactions on Autonomous and
Adaptive Systems*, 11(4):24:1–24:31, 2017. ISSN 1556-4665. doi: 10.1145/3024188. URL
http://doi.acm.org/10.1145/3024188.

75 P. Flach. *Machine Learning: The Art and Science of Algorithms That Make Sense of Data*. Cambridge University Press, 2012. ISBN 9781107422223. URL `https://www.cambridge.org/be/academic/subjects/computer-science/pattern-recognition-and-machine-learning/`.

76 World Economic Forum. How to Prevent Discriminatory Outcomes in Machine Learning, November 2019. URL `https://www.weforum.org/whitepapers/how-to-prevent-discriminatory-outcomes-in-machine-learning/`.

77 E. Fosler-Lussier. Markov Models and Hidden Markov Models: A Brief Tutorial. Berkeley Technical Report TR-98-041, November 2019. URL `http://www.icsi.berkeley.edu/ftp/global/pub/techreports/1998/tr-98-041.pdf`.

78 E. Fredericks, B. DeVries, and B. Cheng. AutoRELAX: Automatically RELAXing a Goal Model to Address Uncertainty. *Empirical Software Engineering*, 19(5):1466–1501, 2014. doi: 10.1007/s10664-014-9305-0. URL `https://doi.org/10.1007/s10664-014-9305-0`.

79 P. Gandodhar and S. Chaware. Context Aware Computing Systems: A Survey. In *2nd International Conference on IoT in Social, Mobile, Analytics and Cloud*, pages 605–608, Tamil Nadu, India, 2018. doi: 10.1109/I-SMAC.2018.8653786. URL `https://ieeexplore.ieee.org/document/8653786`.

80 D. Garlan and B. Schmerl. Model-based Adaptation for Self-healing Systems. In *First Workshop on Self-healing Systems*, pages 27–32, Charleston, South Carolina, 2002. ACM. ISBN 1-58113-609-9. doi: 10.1145/582128.582134. URL `http://doi.acm.org/10.1145/582128.582134`.

81 D. Garlan, S. Cheng, A. Huang, B. Schmerl, and P. Steenkiste. Rainbow: Architecture-Based Self-Adaptation with Reusable Infrastructure. *IEEE Computer*, 37(10):46–54, October 2004. ISSN 0018-9162. doi: 10.1109/MC.2004.175. URL `https://doi.org/10.1109/MC.2004.175`.

82 E. Gat. Three-layer Architectures. In D. Kortenkamp, P. Bonasso, and R. Murphy, editors, *Artificial Intelligence and Mobile Robots: Case Studies of Successful Robot Systems*, pages 195–210. MIT Press, 1998. ISBN 0-262-61137-6. doi: 10.1.1.43.9376. URL `http://dl.acm.org/citation.cfm?id=292092.292130`.

83 D. Gelernter. Generative Communication in Linda. *ACM Transactions on Programming Language Systems*, 7(1):80–112, January 1985. ISSN 0164-0925. doi: 10.1145/2363.2433. URL `http://doi.acm.org/10.1145/2363.2433`.

84 S. Gerasimou, R. Calinescu, S. Shevtsov, and D. Weyns. UNDERSEA: An Exemplar for Engineering Self-Adaptive Unmanned Underwater Vehicles. In *12th International Symposium on Software Engineering for Adaptive and Self-Managing Systems*, pages 83–89, Buenos Aires, Argentina, 2017. doi: 10.1109/SEAMS.2017.19. URL `https://doi.org/10.1109/SEAMS.2017.19`.

85 S. Gerasimou, R. Calinescu, S. Shevtsov, and D. Weyns. UNDERSEA Artifact, November 2019. URL `https://www.hpi.uni-potsdam.de/giese/public/selfadapt/exemplars/undersea/`.

86 S. Gerasimou, R. Calinescu, S. Shevtsov, and D. Weyns. UNDERSEA Website, November 2019. URL `https://www-users.cs.york.ac.uk/simos/UNDERSEA/`.

87 D. Ghosh, R. Sharman, Raghav R., and S. Upadhyaya. Self-healing Systems – Survey and Synthesis. *Decision Support Systems*, 42(4):2164–2185, January 2007. ISSN

0167-9236. doi: 10.1016/j.dss.2006.06.011. URL http://dx.doi.org/10.1016/j.dss.2006.06.011.

88 T. Glazier, B. Schmerl, J. Camara, and D. Garlan. Utility Theory for Self-Adaptive Systems. Carnegie Mellon University Technical Report CMU-ISR-17-119, 2017. URL http://acme.able.cs.cmu.edu/pubs/uploads/pdf/CMU-ISR-17-119.pdf.

89 M. Gorlick and R. Razouk. Using Weaves for Software Construction and Analysis. In *13th International Conference on Software Engineering*, pages 23–34, Austin, Texas, USA, 1991. IEEE. ISBN 0-89791-391-4. doi: 10.1109/ICSE.1991.130620. URL http://dl.acm.org/citation.cfm?id=256664.256677.

90 E. Grua, I. Malavolta, and P. Lago. Self-adaptation in Mobile Apps: A Systematic Literature Study. In *14th International Symposium on Software Engineering for Adaptive and Self-Managing Systems*, pages 51–62, Montreal, Quebec, Canada, 2019. IEEE. doi: 10.1109/SEAMS.2019.00016. URL https://doi.org/10.1109/SEAMS.2019.00016.

91 M. Harman. The Current State and Future of Search Based Software Engineering. In *Future of Software Engineering*, pages 342–357, Minneapolis, MN, USA, 2007. IEEE. ISBN 0-7695-2829-5. doi: 10.1109/FOSE.2007.29. URL https://doi.org/10.1109/FOSE.2007.29.

92 M. Harman. The Role of Artificial Intelligence in Software Engineering. In *Realizing AI Synergies in Software Engineering*, Zurich, Switzerland, 2012. IEEE. doi: 10.1109/RAISE.2012.6227961. URL https://ieeexplore.ieee.org/document/6227961.

93 M. Harman, S. Mansouri, and Y. Zhang. Search-based Software Engineering: Trends, Techniques and Applications. *ACM Computing Surveys*, 45(1):11:1–11:61, December 2012. ISSN 0360-0300. doi: 10.1145/2379776.2379787. URL http://doi.acm.org/10.1145/2379776.2379787.

94 J. Hellerstein, Y. Diao, S. Parekh, and D. Tilbury. *Feedback Control of Computing Systems*. John Wiley Sons, Inc., USA, 2004. ISBN 9780471266372. doi: 10.1002/047166880X. URL https://onlinelibrary.wiley.com/doi/book/10.1002/047166880X.

95 M. Herlihy. Blockchains from a Distributed Computing Perspective. *Communications of the ACM*, 62(2):78–85, January 2019. ISSN 0001-0782. doi: 10.1145/3209623. URL https://doi.org/10.1145/3209623.

96 H. Hermes and J. Lasalle. *Functional Analysis and Time Optimal Control, Volume 56*. Elsevier, 1969. ISBN 9780080955650. URL https://www.elsevier.com/books/functional-analysis-and-time-optimal-control/hermes/978-0-12-342650-5.

97 F. Heylighen. The Science of Self-organization and Adaptivity. In L. D. Kiel, editor, *Knowledge Management, Organizational Intelligence and Learning, and Complexity: Volume III*. EOLSS Publishers Co Ltd, 2002. ISBN 978-1-84826-913-2. URL https://www.eolss.net/.

98 M. Hinchey and R. Sterritt. Self-Managing Software. *Computer*, 39(2): 107–109, February 2006. ISSN 0018-9162. doi: 10.1109/MC.2006.69. URL https://doi.org/10.1109/MC.2006.69.

99 P. Horn. Autonomic Computing: IBM's Perspective on the State of Information Technology. IBM Technical Report, November 2019. URL `https://www.semanticscholar.org/paper/Autonomic-Computing~3A-IBM~27s-Perspective-on-the-State-Horn/1ad1c619a9b3ba5a3ac597f51c8d15011a83423b.`

100 IBM. An Architectural Blueprint for Autonomic Computing, November 2019. URL `https://www-03.ibm.com/autonomic/pdfs/ACBlueprintWhitePaperV7.pdf.`

101 U. Iftikhar and D. Weyns. ActivFORMS: Active Formal Models for Self-adaptation. In *9th International Symposium on Software Engineering for Adaptive and Self-Managing Systems*, pages 125–134, Hyderabad, India, 2014. ACM. ISBN 978-1-4503-2864-7. doi: 10.1145/2593929.2593944. URL `http://doi.acm.org/10.1145/2593929.2593944.`

102 U. Iftikhar, G. Ramachandran, P. Bollansée, D. Weyns, and D. Hughes. DeltaIoT: A Self-adaptive Internet of Things Exemplar. In *12th International Symposium on Software Engineering for Adaptive and Self-Managing Systems*, pages 76–82, Buenos Aires, Argentina, 2017. IEEE. ISBN 978-1-5386-1550-8. doi: 10.1109/SEAMS.2017.21. URL `https://doi.org/10.1109/SEAMS.2017.21.`

103 U. Iftikhar, G. Sankar Ramachandran, P. Bollansee, D. Weyns, and D. Hughes. DeltaIoT Artifact, November 2019. URL `https://www.hpi.uni-potsdam.de/giese/public/selfadapt/exemplars/deltaiot/.`

104 D. Iglesia and D. Weyns. MAPE-K Formal Templates to Rigorously Design Behaviors for Self-Adaptive Systems. *Transactions on Autonomous and Adaptive Systems*, 10(3):15:1–15:31, September 2015. ISSN 1556-4665. doi: 10.1145/2724719. URL `http://doi.acm.org/10.1145/2724719.`

105 ISO/IEC25010. 25010 Standard, November 2019. URL `https://www.iso.org/standard/35733.html.`

106 M. Jackson. The Meaning of Requirements. *Annals of Software Engineering*, 3(1): 5–21, 1997. ISSN 1573-7489. doi: 10.1023/A:1018990005598. URL `https://doi.org/10.1023/A:1018990005598.`

107 P. Jamshidi, A. Sharifloo, C. Pahl, H. Arabnejad, A. Metzger, and G. Estrada. Fuzzy self-learning controllers for elasticity management in dynamic cloud architectures. In *12th International ACM SIGSOFT Conference on Quality of Software Architectures*, pages 70–79, Venice, Italy, 2016. doi: 10.1109/QoSA.2016.13. URL `https://ieeexplore.ieee.org/document/7515437.`

108 P. Jamshidi, J. Camara, B. Schmerl, C. Käestner, and D. Garlan. Machine Learning Meets Quantitative Planning: Enabling Self-Adaptation in Autonomous Robots. In *14th International Symposium on Software Engineering for Adaptive and Self-Managing Systems*, pages 39–50, Montreal, Quebec Canada, 2019. doi: 10.1109/SEAMS.2019.00015. URL `https://dl.acm.org/doi/10.1109/SEAMS.2019.00015.`

109 P. Janert. *Feedback Control for Computer Systems: Introducing Control Theory to Enterprise Programmers*. O'Reilly, 2013. ISBN 9781449361693. URL `http://shop.oreilly.com/product/0636920028970.do.`

110 N. Jennings, P. Faratin, A. Lomuscio, S. Parsons, M. Wooldridge, and C. Sierra. Automated Negotiation: Prospects, Methods and Challenges. *Group Decision and Negotiation*, 10(2):199–215, March 2001. ISSN 1572-9907. doi: 10.1023/A:1008746126376. URL https://doi.org/10.1023/A:1008746126376.

111 I. Jureta, J. Mylopoulos, and S. Faulkner. Revisiting the Core Ontology and Problem in Requirements Engineering. In *2008 16th IEEE International Requirements Engineering Conference*, pages 71–80, Catalunya, Spain, 2008. ISBN 978-0-7695-3309-4. doi: 10.1109/RE.2008.13. URL https://ieeexplore.ieee.org/document/4685655.

112 J. Kephart and D. Chess. The Vision of Autonomic Computing. *IEEE Computer*, 36 (1):41–50, January 2003. ISSN 0018-9162. doi: 10.1109/MC.2003.1160055. URL http://dx.doi.org/10.1109/MC.2003.1160055.

113 J. Kephart and W. Walsh. An Artificial Intelligence Perspective on Autonomic Computing Policies. In *5th IEEE International Workshop on Policies for Distributed Systems and Networks*, Yorktown Heights, NY, USA, 2004. doi: 10.1109/POLICY.2004.1309145. URL https://ieeexplore.ieee.org/document/1309145.

114 N. Khakpour, C. Skandylas, G. S. Nariman, and D. Weyns. Towards Secure Architecture-Based Adaptations. In *14th International Symposium on Software Engineering for Adaptive and Self-Managing Systems*, pages 114–125, Montreal, Canada, 2019. doi: 10.1109/SEAMS.2019.00023. URL https://dl.acm.org/doi/10.1109/SEAMS.2019.00023.

115 G. Kiczales, J. des Rivieres, and Bobrow. D. *The Art of the Metaobject Protocol*. The MIT Press. Cambridge, Massachusetts, 1991. ISBN 9780262111584. URL https://mitpress.mit.edu/books/art-metaobject-protocol.

116 G. Kiczales, J. Lamping, A. Mendhekar, C. Maeda, C. Lopes, Je. Loingtier, and J. Irwin. Aspect-oriented Programming. In *European Conference on Object-Oriented Programming*, pages 220–242, Jyvaskyla, Finland,, 1997. Springer. ISBN 978-3-540-69127-3. doi: 10.1007/BFb0053381. URL https://link.springer.com/chapter/10.1007/BFb0053381.

117 C. Kinneer, Z. Coker, J. Wang, D. Garlan, and C. Le Goues. Managing Uncertainty in Self-adaptive Systems with Plan Reuse and Stochastic Search. In *13th International Symposium on Software Engineering for Adaptive and Self-Managing Systems*, pages 40–50, Gothenburg, Sweden, 2018. ACM. ISBN 978-1-4503-5715-9. doi: 10.1145/3194133.3194145. URL http://doi.acm.org/10.1145/3194133.3194145.

118 B. Kitchenham. Procedures for Performing Systematic Reviews. Keele University Technical Report TR/SE-0401, November 2019. URL http://www.inf.ufsc.br/~aldo.vw/kitchenham.pdf.

119 S. Kounev, J.O. Kephart, A. Milenkoski, and X. Zhu, editors. *Self-Aware Computing Systems*. Springer, 2017. ISBN 978-3-319-47474-8. URL https://www.springer.com/gp/book/9783319474724.

120 J. Kramer and J. Magee. The Evolving Philosophers Problem: Dynamic Change Management. *IEEE Transactions on Software Engineering*, 16 (11):1293–1306, 1990. ISSN 0098-5589. doi: 10.1109/32.60317. URL https://ieeexplore.ieee.org/document/60317.

121 J. Kramer and J. Magee. Self-Managed Systems: An Architectural Challenge. In *Future of Software Engineering. FOSE '07*, pages 259–268, Minneapolis, MN, USA, 2007. doi: 10.1109/FOSE.2007.19. URL `https://ieeexplore.ieee.org/document/4221625`.

122 D. Kusic, J. Kephart, J. Hanson, N. Kandasamy, and G. Jiang. Power and Performance Management of Virtualized Computing Environments via Lookahead Control. *Cluster Computing*, 12(1):1–15, March 2009. ISSN 1386-7857. doi: 10.1007/s10586-008-0070-y. URL `http://dx.doi.org/10.1007/s10586-008-0070-y`.

123 M. Kwiatkowska, G. Norman, and D. Parker. PRISM: Probabilistic Symbolic Model Checker. In T. Field, P. Harrison, J. Bradley, and U. Harder, editors, *Computer Performance Evaluation: Modelling Techniques and Tools*, pages 200–204, London, UK, 2002. Springer. ISBN 978-3-540-46029-9. doi: 10.5555/2944225.2944369. URL `https://link.springer.com/chapter/10.1007/3-540-46029-2_13`.

124 A. Lamsweerde. *Requirements Engineering: From System Goals to UML Models to Software*. Wiley, 2009. ISBN 978-0-470-01270-3. URL `https://www.wiley.com/en-us/Requirements+Engineering\~3A+From+System+Goals+to+UML+Models+to+Software+Specifications-p-9780470012703`.

125 J. Leikas, R. Koivisto, and N. Gotcheva. Ethical Framework for Designing Autonomous Intelligent Systems. *Journal of Open Innovation: Technology, Market, and Complexity*, 5(1), March 2019. doi: 10.3390/joitmc5010018. URL `https://doi.org/10.3390/joitmc5010018`.

126 A. Leva. An Introduction to Systems and Control Theory for Computer Scientists and Engineers. In *8th ACM/SPEC International Conference on Performance Engineering*, pages 433–436, L'Aquila, Italy, 2017. ACM. ISBN 978-1-4503-4404-3. doi: 10.1145/3030207.3053677. URL `http://doi.acm.org/10.1145/3030207.3053677`.

127 W. Levine. *The Control Handbook: Control System Applications, Second Edition*. CRC Press, 2010. ISBN 9781420073607. doi: 10.1201/b10384. URL `https://doi.org/10.1201/b10384`.

128 H. Lim, S. Babu, J. Chase, and S Parekh. Automated Control in Cloud Computing: Challenges and Opportunities. In *Workshop on Automated Control for Datacenters and Clouds*, pages 13–18, Barcelona, Spain, 2009. ACM. ISBN 978-1-60558-585-7. doi: 10.1145/1555271.1555275. URL `http://doi.acm.org/10.1145/1555271.1555275`.

129 P. Maes. Computational Reflection. In K. Morik, editor, *11th German Workshop on Artificial Intelligence*, pages 251–265, Geseke, Germany, 1987. Springer. ISBN 978-3-642-73005-4. URL `https://dl.acm.org/doi/proceedings/10.5555/647607`.

130 P. Maes. *Computational Reflection*. Vrije Universiteit Brussel, Belgium, 1987. URL `http://soft.vub.ac.be/Publications/1987/vub-arti-phd-87_2.pdf`.

131 J. Magee and J. Kramer. Dynamic Structure in Software Architectures. In *4th ACM SIGSOFT Symposium on Foundations of Software Engineering*, pages 3–14, San Francisco, California, USA, 1996. ACM. ISBN 0-89791-797-9. doi: 10.1145/239098.239104. URL `http://doi.acm.org/10.1145/239098.239104`.

132 J. Magee, N. Dulay, S. Eisenbach, and J. Kramer. Specifying Distributed Software Architectures. In *5th European Software Engineering Conference*, London, UK, 1995. Springer. ISBN 3-540-60406-5. doi: 10.1007/3-540-60406-5_12. URL http://dl.acm.org/citation.cfm?id=645385.651497.

133 M. Maggio and H. Hoffmann. ARPE: A Tool To Build Equation Models of Computing Systems. In *8th International Workshop on Feedback Computing*, San Jose, CA, USA, 2013. USENIX Association. URL https://www.usenix.org/node/174699.

134 M. Maggio, A. Papadopoulos, A. Filieri, and H. Hoffmann. Automated Control of Multiple Software Goals Using Multiple Actuators. In *11th Joint Meeting on Foundations of Software Engineering*, pages 373–384, Paderborn, Germany, 2017. ACM. ISBN 978-1-4503-5105-8. doi: 10.1145/3106237.3106247. URL http://doi.acm.org/10.1145/3106237.3106247.

135 M. Maggio, A. V. Papadopoulos, A. Filieri, and H. Hoffmann. Self-Adaptive Video Encoder: Comparison of Multiple Adaptation Strategies Made Simple. In *2017 IEEE/ACM 12th International Symposium on Software Engineering for Adaptive and Self-Managing Systems*, pages 123–128, Buenos Aires, Argentina, 2017. doi: 10.1109/SEAMS.2017.16. URL https://ieeexplore.ieee.org/document/7968140.

136 M. Maggio, A. Vittorio Papadopoulos, A. Filieri, and H. Hoffmann. Self-Adaptive Video Encoder Artifact, November 2019. URL https://www.hpi.uni-potsdam.de/giese/public/selfadapt/exemplars/self-adaptive-video-encoder/.

137 S. Mahdavi-Hezavehi, P. Avgeriou, and D. Weyns. A Classification Framework of Uncertainty in Architecture-Based Self-Adaptive Systems with Multiple Quality Requirements. In I. Mistrik, N. Ali, R. Kazman, J. Grundy, and B. Schmerl, editors, *Managing Trade-Offs in Adaptable Software Architectures*, pages 45–77. Morgan Kaufmann, 2017. ISBN 978-0-12-802855-1. doi: https://doi.org/10.1016/B978-0-12-802855-1.00003-4. URL http://www.sciencedirect.com/science/article/pii/B9780128028551000034.

138 T. Malone, T. Malone, and K. Crowston. The Interdisciplinary Study of Coordination. *ACM Computing Surveys*, 26(1):87–119, March 1994. ISSN 0360-0300. doi: 10.1145/174666.174668. URL http://doi.acm.org/10.1145/174666.174668.

139 D.H. Mellor. *The Facts of Causation*. Routledge. International Library of Philosophy, 1995. ISBN 0-415-09779-7.

140 G. Mohay, E. Ahmed, S. Bhatia, A. Nadarajan, B. Ravindran, A. Tickle, and R. Vijayasarathy. Detection and Mitigation of High-Rate Flooding Attacks. In S. Raghavan and E. Dawson, editors, *An Investigation into the Detection and Mitigation of Denial of Service (DoS) Attacks*. Springer, 2011. ISBN 978-81-322-0276-9. doi: 10.1007/978-81-322-0277-6_5. URL https://link.springer.com/chapter/10.1007/978-81-322-0277-6_5.

141 S. Moon, K. Lee, and D. Lee. Fuzzy Branching Temporal Logic. *Transactions on Systems, Man and Cybernetics, Part B*, 34(2):1045–1055, April 2004. ISSN 1083-4419. doi: 10.1109/TSMCB.2003.819485. URL http://dx.doi.org/10.1109/TSMCB.2003.819485.

142 G. Moreno, B. Schmerl, and D. Garlan. SWIM: An Exemplar for Evaluation and C@bomparison of Self-adaptation Approaches for Web Applications. In *13th International Symposium on Software Engineering for Adaptive and Self-Managing Systems*, pages 137–143, Gothenburg, Sweden, 2018. ACM. ISBN 978-1-4503-5715-9. doi: 10.1145/3194133.3194163. URL http://doi.acm.org/10.1145/3194133 .3194163.

143 G. Moreno, B. Schmerl, and D. Garlan. SWIM Artifact, November 2019. URL https://www.hpi.uni-potsdam.de/giese/public/selfadapt/ exemplars/swim/.

144 G. A. Moreno, J. Cámara, D. Garlan, and B. Schmerl. Proactive Self-adaptation Under Uncertainty: A Probabilistic Model Checking Approach. In *Foundations of Software Engineering*, pages 1–12, Bergamo, Italy, 2015. ACM. ISBN 978-1-4503-3675-8. doi: 10.1145/2786805.2786853. URL http://doi.acm.org/10.1145/2786805 .2786853.

145 B. Morin, O. Barais, J.M. Jezequel, F. Fleurey, and A. Solberg. Models@ Run.time to Support Dynamic Adaptation. *IEEE Computer*, 42 (10):44–51, October 2009. doi: 10.1109/MC.2009.327. URL https://ieeexplore.ieee.org/document/ 5280651.

146 L. Nahabedian, V. Braberman, N. D'ippolito, J. Kramer, and S. Uchitel. Dynamic Reconfiguration of Business Processes. In *International Conference on Business Process Management*, pages 35–51, Vienna, Austria, 2019. Springer. ISBN 978-3-030-26619-6. doi: 10.1007/978-3-030-26619-6_5. URL https://link.springer.com/chapter/ 10.1007/978-3-030-26619-6_5.

147 P. Norvig. Artificial Intelligence in the Software Engineering Workflow. In *O?Reilly Artificial Intelligence Conference*, New York, NY, USA, 2017. URL https://www .oreilly.com/radar/artificial-intelligence-in-the-software- engineering-workflow/.

148 B. Nuseibeh and S. Easterbrook. Requirements Engineering: A Roadmap. In *The Future of Software Engineering*, pages 35–46, Limerick, Ireland, 2000. ACM. ISBN 1-58113-253-0. doi: 10.1145/336512.336523. URL http://doi.acm.org/10.1145/ 336512.336523.

149 IEEE-CS/ACM Joint Task Force on Software Engineering Ethics and Professional Practices. Code of Ethics, November 2019. URL https://www.computer.org/ education/code-of-ethics.

150 P. Oreizy, N. Medvidovic, and R. Taylor. Architecture-based Runtime Software Evolution. In *20th International Conference on Software Engineering*, pages 177–186, Kyoto, Japan, 1998. IEEE. ISBN 0-8186-8368-6. URL http://dl.acm.org/citation .cfm?id=302163.302181.

151 P. Oreizy, M. Gorlick, R. Taylor, D. Heimbigner, G. Johnson, N. Medvidovic, A. Quilici, D. Rosenblum, and A. Wolf. An Architecture-Based Approach to Self-Adaptive Software. *IEEE Intelligent Systems*, 14(3):54–62, May 1999. ISSN 1541-1672. doi: 10.1109/5254.769885. URL http://dx.doi.org/10.1109/5254.769885.

152 P. Padala, K. Shin, X. Zhu, M. Uysal, Z. Wang, S. Singhal, and K. Merchant, A. and Salem. Adaptive Control of Virtualized Resources in Utility Computing Environments. In *2nd ACM SIGOPS/EuroSys European Conference on Computer Systems*, pages 289–302, Lisbon, Portugal, 2007. ACM. ISBN 978-1-59593-636-3. doi: 10.1145/1272996.1273026. URL http://doi.acm.org/10.1145/1272996.1273026.

153 M. Parashar and S. Hariri. Autonomic Computing: An Overview. In J. Banâtre, P. Fradet, J. Giavitto, and O. Michel, editors, *International Workshop on Unconventional Programming Paradigms*, pages 257–269, Le Mont Saint Michel, France, 2005. Springer. ISBN 978-3-540-31482-0. doi: 10.1007/11527800_20. URL https://link.springer.com/chapter/10.1007/11527800_20.

154 C. Perera, A. Zaslavsky, P. Christen, and D. Georgakopoulos. Context Aware Computing for The Internet of Things: A Survey. *IEEE Communications Surveys Tutorials*, 16(1):414–454, First Quarter 2014. ISSN 2373-745X. doi: 10.1109/SURV.2013.042313.00197. URL https://ieeexplore.ieee.org/document/6512846.

155 D. Perez-Palacin and R. Mirandola. Uncertainties in the Modeling of Self-adaptive Systems: A Taxonomy and an Example of Availability Evaluation. In *5th ACM/SPEC International Conference on Performance Engineering*, pages 3–14, Dublin, Ireland, 2014. ACM. ISBN 978-1-4503-2733-6. doi: 10.1145/2568088.2568095. URL http://doi.acm.org/10.1145/2568088.2568095.

156 PRISM. Probabilistic Symbolic Model Checker, November 2019. URL https://www.prismmodelchecker.org/.

157 N. Privault. *Understanding Markov Chains*. Springer, 2018. ISBN 978-981-13-0658-7. doi: https://doi.org/10.1007/978-981-13-0659-4. URL https://link.springer.com/book/10.1007/978-981-13-0659-4#about.

158 M. Provoost and D. Weyns. DingNet: A Self-adaptive Internet-of-Things Exemplar. In *14th International Symposium on Software Engineering for Adaptive and Self-Managing Systems*, Montreal, QC, Canada, 2019. doi: 10.1109/SEAMS.2019.00033. URL https://doi.org/10.1109/SEAMS.2019.00033.

159 F. Quin, D. Weyns, T. Bamelis, S. Buttar, and S. Michiels. Efficient Analysis of Large Adaptation Spaces in Self-adaptive Systems Using Machine Learning. In *14th International Symposium on Software Engineering for Adaptive and Self-Managing Systems*, pages 1–12, Montreal, Quebec, Canada, 2019. IEEE. doi: 10.1109/SEAMS.2019.00011. URL https://doi.org/10.1109/SEAMS.2019.00011.

160 S. V. Raghavan and E. Dawson. *An Investigation into the Detection and Mitigation of Denial of Service (DoS) Attacks: Critical Information Infrastructure Protection*. Springer, 2011. ISBN 978-81-322-0277-6. URL https://www.springer.com/gp/book/9788132202769.

161 V. Rajlich. Software Evolution and Maintenance. In *Future of Software Engineering*, pages 133–144, Hyderabad, India, 2014. ACM. ISBN 978-1-4503-2865-4. doi: 10.1145/2593882.2593893. URL http://doi.acm.org/10.1145/2593882.2593893.

162 G. S. Ramachandran, N. Matthys, W. Daniels, W. Joosen, and D. Hughes. Building Dynamic and Dependable Component-Based Internet-of-Things Applications with

Dawn. In *19th International ACM SIGSOFT Symposium on Component-Based Software Engineering*, pages 97–106, Venice, Italy, 2016. doi: 10.1109/CBSE.2016.18. URL `https://ieeexplore.ieee.org/document/7497436`.

163 A. Ramirez and B. Cheng. Design Patterns for Developing Dynamically Adaptive Systems. In *International Symposium on Software Engineering for Adaptive and Self-Managing Systems*, pages 49–58, Cape Town, South Africa, 2010. ACM. doi: 10.1145/1808984.1808990. URL `https://doi.org/10.1145/1808984.1808990`.

164 A. J. Ramirez, A. Jensen, B. Cheng, and D. Knoester. Automatically Exploring how Uncertainty Impacts Behavior of Dynamically Adaptive Systems. In *26th IEEE/ACM International Conference on Automated Software Engineering*, pages 568–571, Lawrence, KS, USA, 2011. doi: 10.1109/ASE.2011.6100127. URL `https://ieeexplore.ieee.org/document/6100127`.

165 S. Redwine and W. Riddle. Software Technology Maturation. In *8th International Conference on Software Engineering*, London, England, 1985. IEEE. ISBN 0-8186-0620-7. doi: 10.5555/319568.319624. URL `http://dl.acm.org/citation.cfm?id=319568.319624`.

166 N. Rocco De, L. Michele, P. Rosario, and T. Francesco. A Formal Approach to Autonomic Systems Programming: The SCEL Language. *ACM Transactions on Autonomous and Adaptive Systems*, 9(2):7:1–7:29, July 2014. ISSN 1556-4665. doi: 10.1145/2619998. URL `http://doi.acm.org/10.1145/2619998`.

167 R. Rouvoy, P. Barone, Y. Ding, F. Eliassen, S. Hallsteinsen, J. Lorenzo, A. Mamelli, and U. Scholz. MUSIC: Middleware Support for Self-Adaptation in Ubiquitous and Service-Oriented Environments. In B. Cheng, R. de Lemos, H. Giese, P. Inverardi, and J. Magee, editors, *Software Engineering for Self-Adaptive Systems*, pages 164–182. Springer, 2009. ISBN 978-3-642-02161-9. doi: 10.1007/978-3-642-02161-9_9. URL `https://doi.org/10.1007/978-3-642-02161-9_9`.

168 RUBiS. Rice University Bidding System, November 2019. URL `https://www.rubis.ow2.org`.

169 S. Russell and P. Norvig. *Artificial Intelligence: A Modern Approach*. Pearson Education, 2009. ISBN 978-0-13-604259-4. doi: 10.1016/j.artint.2011.01.005. URL `http://aima.cs.berkeley.edu/`.

170 M. Salehie and L. Tahvildari. Self-adaptive Software: Landscape and Research Challenges. *ACM Transactions on Autonomous and Adaptive Systems*, 4 (2):14:1–14:42, May 2009. ISSN 1556-4665. doi: 10.1145/1516533.1516538. URL `http://doi.acm.org/10.1145/1516533.1516538`.

171 P. Sawyer, N. Bencomo, J. Whittle, E. Letier, and A. Finkelstein. Requirements-Aware Systems: A Research Agenda for RE for Self-adaptive Systems. In *18th IEEE International Requirements Engineering Conference*, pages 95–103, Banff, Alberta, Canada, 2010. doi: 10.1109/RE.2010.21. URL `https://ieeexplore.ieee.org/document/5636882`.

172 B. Schilit, N. Adams, and R. Want. Context-Aware Computing Applications. In *1st Workshop on Mobile Computing Systems and Applications*, pages 85–90, Santa Cruz, California, USA, 1994. IEEE. doi: 10.1109/WMCSA.1994.16. URL `https://ieeexplore.ieee.org/document/4624429`.

173 M. Seidl. *UML @ Classroom: An Introduction to Object-Oriented Modeling.* Springer, 2015. ISBN 978-3-319-12742-2. doi: 10.1007/978-3-319-12742-2. URL `https://www.springer.com/gp/book/9783319127415`.

174 M. L. Seto, L. Paull, and S. Saeedi. Introduction to Autonomy for Marine Robots. In M. Seto, editor, *Marine Robot Autonomy*, pages 1–46. Springer, 2013. ISBN 978-1-4614-5659-9. doi: 10.1007/978-1-4614-5659-9_1. URL `https://doi.org/10.1007/978-1-4614-5659-9_1`.

175 M. Shahin, M. Ali Babar, and L. Zhu. Continuous Integration, Delivery and Deployment: A Systematic Review on Approaches, Tools, Challenges and Practices. *IEEE Access*, 5:3909–3943, March 2017. ISSN 2169-3536. doi: 10.1109/ACCESS.2017.2685629. URL `https://ieeexplore.ieee.org/abstract/document/7884954`.

176 S. Shevtsov and D. Weyns. Keep It SIMPLEX: Satisfying Multiple Goals with Guarantees in Control-based Self-adaptive Systems. In *24th ACM SIGSOFT International Symposium on Foundations of Software Engineering*, pages 229–241, Seattle, WA, USA, 2016. ACM. ISBN 978-1-4503-4218-6. doi: 10.1145/2950290.2950301. URL `http://doi.acm.org/10.1145/2950290.2950301`.

177 S. Shevtsov, U. Iftikhar, and D. Weyns. SimCA vs ActivFORMS: Comparing Control- and Architecture-based Adaptation on the TAS Exemplar. In *International Workshop on Control Theory for Software Engineering*, pages 1–8, Bergamo, Italy, 2015. ACM. ISBN 978-1-4503-3814-1. doi: 10.1145/2804337.2804338. URL `http://doi.acm.org/10.1145/2804337.2804338`.

178 S. Shevtsov, M. Berekmeri, D. Weyns, and M. Maggio. Control-Theoretical Software Adaptation: A Systematic Literature Review. *IEEE Transactions on Software Engineering*, 44(8):784–810, Augustus 2018. ISSN 0098-5589. doi: 10.1109/TSE.2017.2704579. URL `https://doi.org/10.1109/TSE.2017.2704579`.

179 S. Shevtsov, D. Weyns, and M. Maggio. SimCA*: A Control-theoretic Approach to Handle Uncertainty in Self-adaptive Systems with Guarantees. *ACM Transactions on Autonomous and Adaptive Systems*, 13(4):17:1–17:34, July 2019. ISSN 1556-4665. doi: 10.1145/3328730. URL `http://doi.acm.org/10.1145/3328730`.

180 S. Shevtsov, D. Weyns., and M. Maggio. Self-Adaptation of Software Using Automatically Generated Control-Theoretical Solutions. In *Engineering Adaptive Software Systems - Communications of NII Shonan Meetings*, pages 35–55. Springer, Kamiyamaguchi, Hayama, Miura District, Kanagawa 240-0198, Japan, 2019. doi: 10.1007/978-981-13-2185-6_2. URL `https://doi.org/10.1007/978-981-13-2185-6_2`.

181 J. Shortle, J. Thompson, D. Gross, and C. Harris. *Fundamentals of Queueing Theory, Fifth Edition.* Wiley, 2018. ISBN 9781118943526. doi: 10.1002/9781119453765. URL `https://onlinelibrary.wiley.com/doi/book/10.1002/9781119453765`.

182 M. Sommer, S. Tomforde, J. Hahner, and D. Auer. Learning a Dynamic Re-combination Strategy of Forecast Techniques at Runtime. In *IEEE International Conference on Autonomic Computing*, pages 261–266, Grenoble, France, 2015. doi: 10.1109/ICAC.2015.70. URL `https://ieeexplore.ieee.org/document/7266977`.

183 V. Souza, A. Lapouchnian, W. Robinson, and J. Mylopoulos. Awareness Requirements for Adaptive Systems. In *6th International Symposium on Software Engineering for*

Adaptive and Self-Managing Systems, pages 60–69, Waikiki, Honolulu, HI, USA, 2011. ACM. ISBN 978-1-4503-0575-4. doi: 10.1145/1988008.1988018. URL `http://doi.acm.org/10.1145/1988008.1988018`.

184 V. Souza, A. Lapouchnian, K. Angelopoulos, and J. Mylopoulos. Requirements-driven Software Evolution. *Computer Science*, 28(4):311–329, November 2013. ISSN 1865-2034. doi: 10.1007/s00450-012-0232-2. URL `http://dx.doi.org/10.1007/s00450-012-0232-2`.

185 G. Tallabaci and V. E. Silva Souza. Engineering Adaptation with Zanshin: An Experience Report. In *8th International Symposium on Software Engineering for Adaptive and Self-Managing Systems*, pages 93–102, San Francisco, CA, USA, 2013. IEEE. doi: 10.1109/SEAMS.2013.6595496. URL `https://ieeexplore.ieee.org/document/6595496`.

186 G. Tesauro and J. Kephart. Utility Functions in Autonomic Systems. In *First International Conference on Autonomic Computing*, pages 70–77, New York, NY, USA, 2004. IEEE. ISBN 0-7695-2114-2. doi: 10.1109/ICAC.2004.68. URL `http://dl.acm.org/citation.cfm?id=1078026.1078411`.

187 R. Thayer and M. Dorfman. *Software Requirements Engineering, 2nd Edition*. Wiley-IEEE Computer Society, 1997. ISBN 978-0-818-67738-0. URL `https://www.wiley.com/en-aw/Software+Requirements+Engineering,+2nd+Edition-p-9780818677380`.

188 H. Tijms. *Stochastic Models, an Algorithmic Approach*. Wiley, 1994. ISBN 978-0471951230.

189 UPPAAL. UPPAAL Tool Suite, November 2019. URL `http://www.uppaal.org/`.

190 H. Van Dyke Parunak, Sven A. Brueckner, and John Sauter. Digital Pheromones for Coordination of Unmanned Vehicles. In *Environments for Multi-Agent Systems*, pages 246–263, New York, USA, 2005. Springer. ISBN 978-3-540-32259-7. URL `https://link.springer.com/chapter/10.1007/978-3-540-32259-7_13`.

191 N. Villegas, H. Müller, G. Tamura, L. Duchien, and R. Casallas. A Framework for Evaluating Quality-driven Self-adaptive Software Systems. In *6th International Symposium on Software Engineering for Adaptive and Self-Managing Systems*, pages 80–89, Waikiki, Honolulu, HI, USA, 2011. ACM. ISBN 978-1-4503-0575-4. doi: 10.1145/1988008.1988020. URL `http://doi.acm.org/10.1145/1988008.1988020`.

192 N. Villegas, G. Tamura, H. Müller, L. Duchien, and R. Casallas. DYNAMICO: A Reference Model for Governing Control Objectives and Context Relevance in Self-Adaptive Software Systems. In R. de Lemos, H. Giese, H. Müller, and M. Shaw, editors, *Software Engineering for Self-Adaptive Systems II*, pages 265–293. Springer, 2013. ISBN 978-3-642-35813-5. doi: 10.1007/978-3-642-35813-5_11. URL `https://doi.org/10.1007/978-3-642-35813-5_11`.

193 T. Vogel. mRUBiS: An Exemplar for Model-based Architectural Self-healing and Self-optimization. In *13th International Conference on Software Engineering for Adaptive and Self-Managing Systems*, pages 101–107, Gothenburg, Sweden, 2018. ACM. ISBN 978-1-4503-5715-9. doi: 10.1145/3194133.3194161. URL `http://doi.acm.org/10.1145/3194133.3194161`.

194 T. Vogel. mRUBIS Artifact, November 2019. URL https://www.hpi.uni-potsdam
.de/giese/public/selfadapt/exemplars/mrubis/.

195 T. Vogel and H. Giese. Model-Driven Engineering of Self-Adaptive Software with
EUREMA. *ACM Transactions on Autonomous and Adaptive Systems*, 8(4):18:1–18:33,
January 2014. ISSN 1556-4665. doi: 10.1145/2555612. URL http://doi.acm.org/
10.1145/2555612.

196 T. Vogel and H. Giese. Self-Adaptive Systems Artifacts and Model Problems, November
2019. URL https://www.hpi.uni-potsdam.de/giese/public/selfadapt/
exemplars/.

197 G. Welch and G. Bishop. An Introduction to the Kalman Filter.
University of North Carolina at Chapel Hill, NC, USA, 1995. URL
https://www.cs.unc.edu/~welch/media/pdf/kalman_intro.pdf.

198 D. Weyns. Software Engineering of Self-adaptive Systems. In S. Cha, R. Taylor, and K.
Kang, editors, *Handbook of Software Engineering*, pages 399–443. Springer, 2019. doi:
10.1007/978-3-030-00262-6_11. URL https://doi.org/10.1007/978-3-030-
00262-6_11.

199 D. Weyns and T. Ahmad. *Claims and Evidence for Architecture-Based Self-adaptation:
A Systematic Literature Review*, pages 249–265. Springer, Montpellier, France, 2013. doi:
10.1007/978-3-642-39031-9_22. URL https://link.springer.com/chapter/10
.1007/978-3-642-39031-9_22.

200 D. Weyns and R. Calinescu. Tele Assistance: A Self-adaptive Service-based System
Examplar. In *10th International Symposium on Software Engineering for Adaptive and
Self-Managing Systems*, pages 88–92, Florence, Italy, 2015. IEEE. URL http://dl
.acm.org/citation.cfm?id=2821357.2821373.

201 D. Weyns and R. Calinescu. TAS Artifact, November 2019. URL https://www.hpi
.uni-potsdam.de/giese/public/selfadapt/exemplars/tas/.

202 D. Weyns and U. Iftikhar. ActivFORMS: A Model-Based Approach to Engineer
Self-Adaptive Systems. *arXiv 1908.11179, cs.SE*, 2019. URL https://arxiv.org/
abs/1908.11179.

203 D. Weyns and M. Provoost. DingNet Website, November 2019. URL
https://people.cs.kuleuven.be/~danny.weyns/software/DingNet/
index.htm.

204 D. Weyns, A. Helleboogh, and T. Holvoet. The Packet-World: A Test Bed for Investi-
gating Situated Multi-Agent Systems. In R. Unland, M. Calisti, and M. Klusch, editors,
Software Agent-Based Applications, Platforms and Development Kits, pages 383–408.
Birkhäuser, 2005. ISBN 978-3-7643-7348-1. URL https://lirias.kuleuven.be/
retrieve/5983.

205 D. Weyns, U. Iftikhar, D. de la Iglesia, and T. Ahmad. A Survey of Formal Methods
in Self-adaptive Systems. In *Fifth International C* Conference on Computer Science
and Software Engineering*, pages 67–79, Montreal, Quebec, Canada, 2012. ACM. ISBN
978-1-4503-1084-0. doi: 10.1145/2347583.2347592. URL http://doi.acm.org/10
.1145/2347583.2347592.

206 D. Weyns, S. Malek, and J. Andersson. FORMS: Unifying Reference Model for Formal
Specification of Distributed Self-adaptive Systems. *Transactions on Autonomous and*

Adaptive Systems, 7(1):8:1–8:61, 2012. ISSN 1556-4665. doi: 10.1145/2168260.2168268. URL http://doi.acm.org/10.1145/2168260.2168268.

207 D. Weyns, S. Malek, J. Andersson, and B. Schmerl. Introduction to the Special Issue on State of the Art in Engineering Self-adaptive Systems. *Journal of Systems and Software*, 85(12):2675–2677, 2012. ISSN 0164-1212. doi: 10.1016/j.jss.2012.07.045. URL https://doi.org/10.1016/j.jss.2012.07.045.

208 D. Weyns, U. Iftikhar, and J. Soderland. Do External Feedback Loops Improve the Design of Self-adaptive Systems? A Controlled Experiment. In *International Symposium on Software Engineering for Adaptive and Self-Managing Systems*, San Francisco, CA, USA, 2013. ISBN 978-1-4673-4401-2. doi: 0.1109/SEAMS.2013.6595487. URL http://dl.acm.org/citation.cfm?id=2487336.2487341.

209 D. Weyns, B. Schmerl, V. Grassi, S. Malek, R. Mirandola, C. Prehofer, J. Wuttke, J. Andersson, H. Giese, and K. Göschka. On Patterns for Decentralized Control in Self-Adaptive Systems. In R. de Lemos, H. Giese, H. Müller, and M. Shaw, editors, *Software Engineering for Self-Adaptive Systems II*, pages 76–107. Springer, 2013. ISBN 978-3-642-35813-5. doi: 10.1007/978-3-642-35813-5_4. URL https://doi.org/10.1007/978-3-642-35813-5_4.

210 D. Weyns, N. Bencomo, R. Calinescu, J. Camara, C. Ghezzi, V. Grassi, L. Grunske, P. Inverardi, J. Jezequel, S. Malek, R. Mirandola, M. Mori, and G. Tamburrelli. Perpetual Assurances for Self-Adaptive Systems. In R. de Lemos, D. Garlan, C. Ghezzi, and H. Giese, editors, *Software Engineering for Self-Adaptive Systems III. Assurances*, pages 31–63. Springer, 2017. ISBN 978-3-319-74183-3. doi: 10.1007/978-3-319-74183-3_2. URL https://link.springer.com/chapter/10.1007/978-3-319-74183-3_2.

211 D. Weyns, U. Iftikhar, D. Hughes, and N. Matthys. Applying Architecture-Based Adaptation to Automate the Management of Internet-of-Things. In *European Conference on Software Architecture*, pages 49–67, Madrid, Spain, 2018. Springer. ISBN 978-3-030-00761-4. doi: 10.1007/978-3-030-00761-4_4. URL https://link.springer.com/chapter/10.1007/978-3-030-00761-4_4.

212 D. Weyns, R. Calinescu, Iftikhar U., and Y. Ruan. TAS Website, November 2019. URL https://people.cs.kuleuven.be/~danny.weyns/software/TAS/.

213 D. Weyns, U. Iftikhar, and G. Sankar Ramachandran. DeltaIoT Website, November 2019. URL https://people.cs.kuleuven.be/~danny.weyns/software/DeltaIoT/.

214 J. Whittle, P. Sawyer, N. Bencomo, B. Cheng, and J. Bruel. RELAX: Incorporating Uncertainty into the Specification of Self-Adaptive Systems. In *17th IEEE International Requirements Engineering Conference*, pages 79–88, Atlanta, Georgia, USA, 2009. IEEE. ISBN 978-0-7695-3761-0. doi: 10.1109/RE.2009.36. URL http://dx.doi.org/10.1109/RE.2009.36.

215 M. Wooldrige. *An Introduction to MultiAgent Systems*. Wiley, USA, 2009. ISBN 978-0-470-51946-2. URL https://www.wiley.com/en-us/An+Introduction+to+MultiAgent+Systems\~2C+2nd+Edition-p-9780470519462.

216 S. Yang. Cybersecurity Threats – Can we Predict Them? In *Research Features Magazine: Engineering and Technology*. Research Publishing International Ltd, 2018. URL https://cdn2.researchfeatures.com/wp-content/uploads/2018/07/Shanchieh-Jay-Yang-1.pdf.

217 E. Yuan, N. Esfahani, and S. Malek. A Systematic Survey of Self-Protecting Software Systems. *ACM Transactions on Autonomous and Adaptive Systems*, 8(4):17:1–17:41, Januari 2014. ISSN 1556-4665. doi: 10.1145/2555611. URL http://doi.acm.org/10.1145/2555611.

218 J. Zhang and B. Cheng. Using Temporal Logic to Specify Adaptive Program Semantics. *Journal of Systems and Software*, 79(10):1361–1369, 2006. ISSN 0164-1212. doi: https://doi.org/10.1016/j.jss.2006.02.062. URL http://www.sciencedirect.com/science/article/pii/S0164121206001397.

219 J. Zhang and B. Cheng. Model-based Development of Dynamically Adaptive Software. In *28th International Conference on Software Engineering*, pages 371–380, Shanghai, China, 2006. ACM. ISBN 1-59593-375-1. doi: 10.1145/1134285.1134337. URL http://doi.acm.org/10.1145/1134285.1134337.

220 T. Zhao, W. Zhang, H. Zhao, and Z. Jin. A Reinforcement Learning-Based Framework for the Generation and Evolution of Adaptation Rules. In *IEEE International Conference on Autonomic Computing*, pages 103–112, Columbus, OH, USA, 2017. doi: 10.1109/ICAC.2017.47. URL https://ieeexplore.ieee.org/document/8005338.

221 F. Zhou, B. Wu, and Z. Li. Deep Meta-Learning: Learning to Learn in the Concept Space. *CoRR*, abs/1802.03596, 2018. URL http://arxiv.org/abs/1802.03596.

Index

a

adaptation action   51
adaptation goals   8, 45, 70, 78, 115, 121
adaptation options   48, 102, 141, 150, 208
adaptation space   48, 151
  reduction   208
adaptive control   177
analyzer   47
  analysis mechanism   49
  basic workflow   47
  tactics   156
architectural model   72
  adapatation strategies   72
    adapation operators   73
  analyses   72
  component model   72
  constraints   72
  properties   72
architecture-based adaptation   18, 63
aspect-oriented programming   65
automata   99, 129, 151
automatic controller construction   177
  formal guarantees   181
  MIMO controllers   184
  model update   179
    incremental   180
    rebuilding   180
  MPC   189
  phases   178
    controller creation   179
    model building   178
    operation   179
  push-buttom methodology   178

  SISO controllers   178
automating tasks   18
  manageability problems   33
autonomic computing   34
autonomous system   4
awareness requirements   123
  types   123

b

Bayesian estimation   204
  transition probability matrix   204
  updating rule   206

c

causality   90
  weak causality   91
classification (learning)   208
  incremental classifier   210
Cloud infrastructure   213
  auto-scaling   214
  scaling rules   215
comprehensive reference model   75
  distribution perspective   79
  MAPE-K perspective   78
  reflection perspective   76
computational reflection   64
context-awareness   4
control theory   171
  basic feedback control loop   173
control-based software adaptation   20
coordination mechanism   81
  coordination channel   82
  coordination model   81
  coordination protocol   81

An Introduction to Self-Adaptive Systems: A Contemporary Software Engineering Perspective,
First Edition. Danny Weyns.
© 2021 John Wiley & Sons Ltd. Published 2021 by John Wiley & Sons Ltd.

d

DeltaIoT   25
  effectors   28
  management interface   27
  multi-hop communication   25
  over provisioning   29
  quality requirements   29
  queues   27
  signal to noice ratio   28
  time synchronized communication   27
  uncertainties   28
distribution versus decentralization   82

e

effector   51
environment   5
essential maintenance tasks   37
evolution management   53
evolution requirements   124
  operators   125
executor   51
  basic workflow   51
external adaptation mechanism   64

f

fading boundaries   137
fail-safe strategy   49
feedback control loop   173
  discrete time system model   174
  PI control   174
  properties   175
    accuracy   175
    overshoot   176
    robustness   176
    settling time   176
    stability   175
  purposes   174
  system model   174
  transfer function   181
    pole   182
  Z-transform   181
feedback loop   8, 34
  functional requirements   127
    deploy and execute   130
    design and verify   128

fuzzy logic controller   214
fuzzy Q-learning   217
  Q-table   218

g

guarantees under uncertainty   20

i

integration evolution and adaptation
    management   55
internal adaptation mechanism   64
Internet-of-Things   25

k

Kalman filter   180, 191
knowledge   44
  adaptation goals   45
  environment model   45
  managed system model   45
  MAPE working model   45

l

learning from experience   20
look-ahead horizon   154

m

machine learning   201
managed system   7
managing system
  primary functions   43
  reference model   44
MAPE-K model   44
  workflow   202
Markov model   96, 144
  DTMC   146, 204
  Markov property   204
  MDP   156
  PCTL   147
meta-object protocol   65
meta-requirements   122
  awareness requirements   123
    semantics   124
  evolution requirements   124
  operationalization   126
model-driven engineering   89

monitor   46
  basic workflow   46
multi-agent system   4

p
plan   51
planner   49
  basic workflow   50
planning mechanism   51
PRISM   147
probe   46

q
quality model   151
queuing model   97
quiescence   7

r
rationale for architectural prespective   64
  abstraction to manage system change   66
  dealing with system-wide concerns   66
  falicitating scalabily   66
  integrated approach   65
  leveraging consolidated efforts   65
  separating domain and adaptation
     concerns   65
reference model of managing system   43
reinforcement learning   217
relaxing requirements   116
  handling uncertainty   118
  language operators   116
  operationalization   118
    mitigation mechanims   119
    requirements reflection   119
  semantics   118
requirements engineering   115
  goal-based modeling   119
requirements-driven adaptation   19
reward/cost structure   156
RUBiS   155
runtime model   90
  definition   90
runtime models   18
  declarative versus procedural   94
  dimensions   92

formal versus informal   98
functional versus qualitative   95
  functional models   95
  quality models   95
MAPE components exchange K models
    103
MAPE components share K models   101
MAPE models share K models   105
motivations   91
principal strategies   101
structural versus behavioral   93
runtime quantitative verification   144

s
self-adaptation   1
  comprehensive reference model   75
  conceptual model   5
  external principle   3
  fading boundaries   17
  internal principle   3
  motivation   33
  seven waves   17
self-adaptation management   54
self-awareness   4
self-configuration   42
self-healing   38
self-optimization   37
self-organizing system   4
self-protection   41
service-based health assistance system   1
seven waves   18
  enabled contributions   20
  schematic overview   18
  selected work   20
simplex method   185
software evolution   53
software-intensive system   1
state-space explosion   149
statistical model checking   152

t
taming uncertainty   137
  analysis adaptation options   141
  exhaustive verification   144
  integrated process   160

taming uncertainty (*contd.*)
    deploy managing system   162
    evolve goals and managing system   163
    four stages   160
    implement and verify managing system
      161
    verify options, decide and adapt   163
  proactive latency-aware adaptation   154
  runtime quantitative verification   144
  selection best adaptation option   143
  statistical model checking   149
three-layer model   66
  change management   67
  component control   67
  goal management   68
  mapping to conceptual model   70

u

uncertainty   1, 139
  sources   139
    context   141
    goals   140
    human involvement   141
    system itself   139
  taming uncertainty   141
  working definition   139
Upaal   152
utility function   36, 146, 156
  expected utility   36